Environmental Due Diligence

Environmental Due Diligence

The Role of ISO 14015 in the Environmental Assessment of Sites and Organizations

Nigel Carter
Larraine Wilde

Business Information

Throughout the text, several companies are named in examples and case studies. These companies are mentioned for illustrative purposes only and their citing is not to be taken as an endorsement by BSI of the companies named.

The authors, Nigel Carter and Larraine Wilde, assert their Moral Rights to be identified as the authors of this work in accordance with Sections 77 and 78 of the Copyright, Designs and Patents Act 1988.

ISBN 0 580 44296 9
BSI Reference: BIP 2038

First published, December 2004

© British Standards Institution 2004.

Copyright subsists in this publication. Except as permitted under the Copyright, Designs and Patents Act 1988, no extract may be reproduced, stored in a retrieval system or transmitted in any form or by any means – electronic, photocopying, recording or otherwise – without prior written permission from BSI. If permission is granted, the terms may include royalty payments or a licensing agreement. Details and advice can be obtained from the Copyright Deparment, BSI, 389 Chiswick High Road, London W4 4AL, UK, copyright@bsi-global.com.

Great care has been taken to ensure accuracy in the compilation and preparation of this publication. However, since it is intended as a guide and not a definitive statement, the authors and BSI cannot in any circumstances accept responsibility for the results of any action taken on the basis of the information contained in the publication nor for any errors or omissions. This does not affect your statutory rights.

Typeset by Monolith – www.monolith.uk.com
Printed by Greenaways

Contents

About the authors ix
Foreword xi
Introduction xiii

Chapter 1 – What is environmental assessment?   1
Introduction   1
Environmental assessment   1
Conclusion   6

Chapter 2 – The client/assessor relationship   7
Introduction   7
The client's role and responsibilities   7
The representative of the assessee   10
The assessor   10
Assessor qualifications   12
Conclusion   13

Chapter 3 – Assessment planning   15
Introduction   15
Objectives and scope   15
Assessment criteria   18
Assessment plan   20
Summary   21

Chapter 4 – The assessment process: information gathering   23
Introduction   23
Information   23
Conclusion   31

Chapter 5 – The assessment process: interviewing and validation   33
Introduction   33
Interviewing   33
Information validation   35
Conclusion   35

Chapter 6 – The assessment process: evaluation of issues and determination of business consequences — 37
Introduction — 37
Evaluation — 37
Determination of business consequences — 38
Conclusion — 39

Chapter 7 – The assessment process: reporting to the client — 41
Introduction — 41
Report contents — 41
Conclusions — 43

Chapter 8 – Intrusive investigation — 45
Sites being decommissioned — 46
Investigations for development sites — 48
Intrusive investigation (Phase 2 investigation) — 50
Other investigations — 52
Conclusions — 53

Chapter 9 – Risk assessment and remediation — 55
Risk assessment — 55
Remediation — 57
Conclusions — 61

Chapter 10 – Emerging legislation — 63
The Liability Directive — 63
Other legislation — 65
Conclusions — 65

Appendix 1 – Remediation technology — 67
Introduction — 67
Contaminant — 68
Depth and extent of contamination — 68
Soil quality — 69
Geographical/physical location — 69
Meterorological conditions — 69
Geological/hydrogeological conditions — 69
Potential hazards from the contaminant and potential contaminant pathways — 70
Groundwater and groundwater extraction — 70

Excavation and disposal	71
Engineering/containment	72
References	**73**

About the authors

Nigel Carter

Following seven years service in the Royal Navy's Executive Branch, Nigel Carter joined BP in 1970, where he worked in general management in marketing and distribution before moving to the Shetlands to specialize in crude oil exploration and production. After a short period in Zambia as the general manager of the country's strategic oil storage installation, Nigel joined BP's corporate headquarters in London in 1983, where he developed a trading desk that ultimately turned over US$22 million per annum, before becoming the project manager for African associates in BP's worldwide re-imaging campaign in 1989.

In 1993, Nigel became self-employed as an environmental management consultant, providing advice on industrial waste management, environmental management and auditing, environmental assessment, corporate reporting and associated skills.

Nigel is a member of Kennet District Council, the Council of Swindon Chamber of Commerce and Industry, and is also the Chairman of BSI Technical Committee SES/1/-/5, *Greenhouse gas management*, which develops standards for the preparation and reporting of greenhouse gas inventories.

Larraine Wilde

Larraine Wilde is an environmental scientist and a senior project manager with Sinclair Knight Merz (Europe). Larraine has undertaken environmental assessments, including due diligence audits, in more than 20 countries in Europe, Asia, South America and Africa, predominantly in the mining and mineral processing sectors. More recently, Larraine led a team of scientists assessing claims of more than US$40 billion brought against Iraq for environmental liabilities resulting from the invasion of Kuwait.

Larraine is Chairman of the BSI Technical Committee SES/1/2, *Environmental auditing and related environmental investigations*, and is also a member of the ISO Technical Committee ISO/TC 207/SC2, *Environmental auditing and related environmental investigations*, which produced environmental audit standards in the ISO 14000 series, including ISO14015.

Foreword

The requirements for environmental due diligence and environmental assessment are prevalent in economies where the free movement of capital is encouraged and grow almost daily. In most cases, the client is 'divorced' from the work and the environmental consultant undertaking the 'due diligence' by an intermediary, typically a solicitor. The client and, frequently, the intermediary usually do not understand the intricacies of the risk assessments being undertaken, particularly where intrusive investigation has been undertaken and remediation recommended. On the other hand, in forming any opinions on 'business consequences', the consultant may not be aware of the commercial intentions of the client. This arrangement does not necessarily lead to an effective outcome for the client, or the most insightful investigation by the consultant.

This book does not set out to provide any in-depth technical guidance on the investigative processes, or unravel the mysteries of the many comprehensive software programmes that lie at the heart of risk assessments. It is intended to be a layman's guide to the processes behind environmental assessment, and an insight into the identification of 'business consequences'. In this context, ISO 14015 forms an excellent guide to the requirements for an initial assessment – often described as a 'first phase' or 'desktop' assessment. This book also discusses the growing impact of legislation on industry, covering the extremes of legislation aimed at the traditionally heavily polluting industries, as well as reference to the Statutory Nuisance legislation, requiring prevention of excessive noise, vibration, odour, dust and litter.

Finally, the book is intended to provide the reader, whether they be a potential client or a legal professional, with a deeper understanding of the processes, so that they are better informed when engaging environmental professionals in both the briefing for the 'due diligence', and when developing the 'business consequences' on which the entire process may depend.

Introduction

For many years, where the ownership of business and property was exchanged, the concept of due diligence was largely confined to an examination of accounts and title deeds, to a review of the client base and the discrete potential for exploitation of assets in general. With the acquisition and disposal of assets and the free movement of capital in many increasingly regulated and litigious international markets, due diligence has now been extended to cover an examination for potential liabilities arising from health, safety and especially environmental impacts. Typically, leadership is taken by law firms that are advising on the acquisition or disposal. While frequently acting as the 'client', they appear to be constrained by their lack of detailed understanding of the processes and risk assessments undertaken. This inherent weakness may explain, in part, the reasons for the failures reported in a survey of UK companies that was conducted by KPMG in December 2002; 80% of the respondents had undertaken environmental due diligence but a significant minority had identified post-transaction issues, almost half of which had been outside the scope of due diligence.

Environmental due diligence is a service provided by many, mostly large, environmental consultancies. Indeed, a search on the Internet using the key words 'environmental due diligence' will reveal a wealth of offerings – case studies, electronic surveys and pro forma studies in a variety of forms – from such consultancies, who offer a presence in most of the markets where there is environmental regulation and the free movement of capital. However, the alert reader will note that many of the company profiles promote environmental due diligence in the context of contaminated land and asbestos. This is, of course, perfectly understandable, given the potential earnings to be derived. At the beginning of the 1990s, one oil company noted that the potential cost of cleaning up of redundant assets – filling/service stations, distribution terminals and refineries – was likely to be in excess of US$100 million. The current cost of remediating severe hydrocarbon and associated contamination at old town gas works may exceed £5 million.

This potential for remediation is supplemented by the progressive de-nationalization of many former parastatal organizations in eastern and central Europe, as well as the reclamation for civilian use of many former military installations.

However, with increasing public awareness of environmental issues and the prospect of increasingly scare access to some, especially strategic, materials, appeal in the market for some products and access to essential raw materials for the manufacture of others gives environmental due diligence a significance beyond the realms of contaminated land.

Finally, one should not ignore the plethora of environmental regulations and the requirement for compliance – elements of these regulations impose not just corporate but potentially personal liability which, in the UK, may extend to fines of up to £20 000 and/or imprisonment for a period up to six months. Exposure to such penalties extends beyond the

Environmental Due Diligence

chief executive or chief operating officer and into the realms of the functional manager with environmental management responsibility.

It should also be mentioned that the responsibility for environmental due diligence does not always fall on potential acquirers or disposers and their legal advisers. Receivers in bankruptcy and, under recent legislation, local government authorities, may find themselves responsible for some potential environmental liabilities for businesses or premises. The concept of the 'orphan' site is now accepted in regulation along with the requirement for the public purse to pay for remediation where the historic polluter cannot be identified.

The overall significance of the environmental due diligence process is to ensure that, in acquiring or disposing of assets, no residual environmental issue arises which might have a potential to significantly reduce the financial benefit from the transaction or call the reputation or corporate governance of the organization into question. This risk may increasingly be offset by the introduction of an environmental indemnity, effectively insuring the acquirer of an asset with recourse either to the seller or, more typically, the insurer offering the indemnity.

It is, finally, important to recognize the role of environmental assessment in relation to other environmental disciplines – environmental impact assessment (EIA), initial environmental review (IER) and environmental auditing. These terms are frequently confused by the layman, and some explanation might now be appropriate. The relationship can best be illustrated in Figure 1.

An environmental impact assessment almost always precedes the development of a project, evaluating both positive and negative impacts of the proposed project against a variety of criteria. These include: reference to biodiversity, archaeological and other historic factors; the impact of any potential emissions to air, water or soil; changes required in the local infrastructure; and provisions for employees'. It is a legislative requirement for all developments involving processes permitted under Integrated Pollution Prevention and Control (IPPC), including the construction of marinas and other similar leisure facilities, airports, electrical generation facilities and other 'big polluters'. Implicit in the process is a requirement to re-visit the impact assessment to corroborate the initial assumptions and expectations with actual performance. This may, of course, to some extent be achieved through the establishment of appropriate objectives in any subsequent implementation of an environmental management system.

An IER is a typical early process in the development of an environmental management system and is referred to in ISO 14004, 4.1.3 [1] as the 'current position of an organization with regard to the environment'. An IER entails a holistic review of the business, identifying relevant legislation, its positive and negative environmental aspects, existing performance in regard to environmental protection and pollution prevention, and current, relevant policies on environmental concerns and the perception of the organization's environmental footprint by third parties. Once an environmental management system is implemented and developed, there is little need for another environmental review.

Introduction

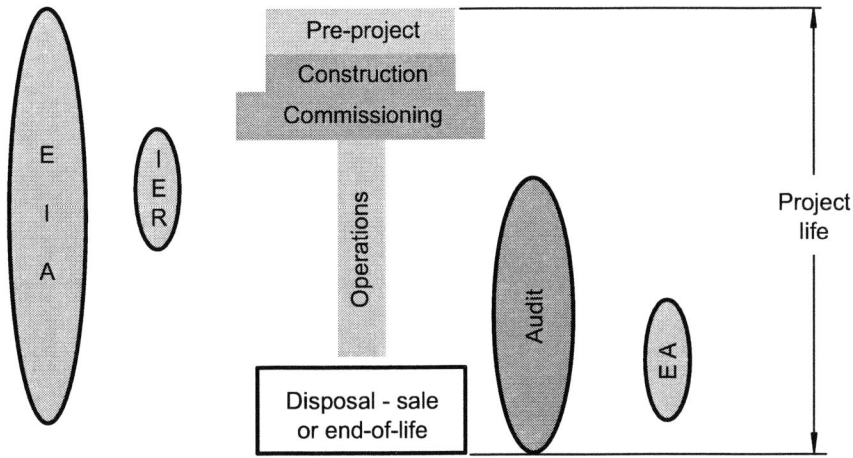

Figure 1 – Relationship between an environmental impact assessment (EIA), initial environmental review (IER), environmental audit and environmental assessment (EA) over a project's life

An audit is defined in ISO 19011, 3.1 [2] as a 'systematic, independent and documented process for obtaining audit evidence and evaluating it objectively to determine the extent to which the audit criteria are fulfilled'. Implicit in this definition is a rigorous process, capable of deriving objective views on the suitability and effectiveness of the management system. It is also intended as a repetitive activity, designed to deliver continual improvement to the management system processes.

This publication is intended to introduce the reader to, in particular, the international standard ISO 14015, *Environmental management – Environmental assessment of sites and organisations* [3], and its application to the non-intrusive stages of environmental assessment. It will conclude with comments on the additional guidance to be obtained from national standards and other sources of intrusive investigation, where this is appropriate.

Chapter 1 – What is environmental assessment?

Introduction

The challenges of contaminated land, compliance with regulation and product liability are becoming increasingly complex and it is vital for businesses to understand the consequences that these may have, not just on financial viability and the relationships with financial stakeholders, but with client perceptions and continuity in production. This chapter covers ISO 14015, Introduction and Sections 1 and 2. ISO 14015 was written with the intention of enabling environmental assessment to be undertaken by a variety of organizations, both large and small. Environmental assessment is defined as the:

> 'process to identify objectively the environmental aspects, to identify the environmental issues and to determine the business consequences of sites and organizations as a result of past, current and expected future activities.'
> (Source: ISO 14015, 2.7)

A note qualifies this definition by saying that the subsequent determination of business consequences is optional and at the discretion of the client. However, it would be commercially unrealistic to anticipate that many organizations would embark on this process without ultimately intending to derive an indication of the financial implications of polluted assets, marginal or non-compliance and/or a production process threatened by reducing access to essential raw materials or by the unsustainability in use of its produce.

In addressing these concerns, the standard expects to contribute to the understanding of environmental aspects that have also been identified for the purposes of implementing an environmental management system. The activities described in ISO 14015 are helpful in fulfilling some of the requirements of an initial environmental review. It would be equally true to say that information derived from environmental audits (for compliance or for conformity with an environmental management systems (EMS)), environmental impact assessments and environmental performance evaluations can contribute significantly to the accumulation of information relating to, especially, current and projected activities.

Environmental assessment

In introducing the concept of environmental assessment, the topic of environmental issues has been discussed. The standard (ISO 14015, 2.9) defines an environmental issue as that 'for which validated information on environmental aspects deviates from the selected criteria and may result in liabilities or benefits on the assessee's or the client's public image, or other costs'. In the somewhat stilted terms of an international standard, the intention here is perhaps a little obscure and examples of the sort of issue that might arise could include:

- the identification of components or raw materials purchased from unethical sources. Examples here might be material produced from child labour or timber from unsustainably managed forests;
- a history of persistent breaches of legislation, whether environmental or not, which would give rise to a loss of the 'license to operate' within the local community. This would portend possibility difficulties in obtaining consent for (re)development or recruiting;
- the use of strategic raw materials to which there is some risk of supply constraint or total loss of access.

Types of environmental issue fall into several categories with the potential to generate liabilities, particularly 'hidden' liabilities that may not become apparent until long after the purchase or that require significant costs to address. These include:

- soil and groundwater contamination;
- human health issues;
- potential failures of large structures such as dams;
- the potential or actual release of hazardous materials and/or wastes;
- the need for significant investment to upgrade processes or add abatement equipment;
- product streams that are reaching the end of their acceptability;
- social impacts such as those described earlier. (It is becoming increasingly difficult to separate social and environmental impacts, particularly for some large projects that require resettlement of communities.)

This is not to say that an acquisition will not go ahead. The purpose of the environmental due diligence (EDD) is to assist the purchaser in calculating the level of financial or other risk associated with the acquisition or activity rather than assessing environmental performance for its own sake. This might be achieved by the submission to the vendor of some form of enquiry but this, demonstrably, has limitations.

The business consequences of pursuing an acquisition, or persisting with current production after such discoveries, include reduced or interrupted production while alternative materials are sourced, a loss of public goodwill and a loss of preference for the company's products. These in turn will affect the company's cash flow and profits, and its potential market value. The discovery, in any subsequent intrusive investigation, of the presence of asbestos or contaminated land will further reduce potential marketability of the asset pending expenditure on remediation (see Figure 2).

Of course, any organization preparing itself for a market flotation or outright sale would benefit from early knowledge of such problems and the ability to seek alternative raw material or component sources, adjust production techniques or pay for remediation from cash flow as opposed to seeing their market valuation reduced at the moment of sale. Surprises for the seller offer the potential acquirer a stronger negotiating position. In utilizing the discoveries from the environmental assessment, it is important to understand what is meant by 'business consequences'. The standard (ISO 14015, 2.3) defines these

What is environmental assessment?

as the 'actual or potential impact (financial or other; positive or negative; qualitative or quantitative) of the identified and evaluated environmental issues'. It is, of course, desirable that any evaluation that is made, and any conclusions that are drawn, should be made on the basis of as much objective evidence as is available.

Figure 2 – Redundant hospital premises on urban fringes

However, as will become apparent as the process is revealed, it is expected that the environmental assessor will be required to exercise perhaps considerable professional judgement in the formation of opinions on potential business consequences. This is perhaps a more significant point than might, at first, be assumed. Undoubtedly, the environmental assessor requires a comprehensive knowledge and understanding of environmental regulation, potentially polluting emissions from industrial processes and the industrial context in which some land contamination may occur. However, this knowledge must be complemented by an appreciation of broader commercial practice, financial values and, probably, the interaction of sector-specific markets. It should be noted that there are no specific professional qualifications for an environmental assessor nor, indeed, any professional institution addressing professional standards for the role. However, professional standing may be ascertained by virtue of membership of one or more of a number of professional organizations – the Institute of Environmental Management and Assessment (IEMA), the Engineering Employers' Federation (EEF) and the Royal Society of Chemists among others. Environmental consultancies may not be necessarily achieving the breadth of analysis that the client is seeking. Meanwhile, the standard identifies (ISO 14015, 2.2) the assessor as a 'person, possessing sufficient competence, designated to conduct or participate in a given assessment'.

The standard also defines (ISO 14015, 2.1 and 2.4) the assessee as the 'site or organization being assessed and the client as the organization commissioning the assessment'. The client may be:

- a law company representing the interests of and advising the potential disposer or acquirer of the asset;
- a financial institution providing a loan;
- a corporate entity acquiring the site or organization.

However, it is also perfectly acceptable for commercial organizations to commission directly an environmental assessment on their own sites and internal activities or on external assets. It should also be pointed out that not all assessments that form part of an environmental due diligence study are necessarily conducted in an overt manner. Indeed, the execution of the assessment may be undertaken in extremely secret or confidential circumstances. These may be required to avoid alerting the owners of potential acquisitions as to the interest their asset is attracting, with consequential concerns over price negotiations. Early recognition of business consequences arising from an environmental issue may be a useful negotiating tool. The roles of the client and assessor are discussed in more detail in the following chapter. However, the significant issue for the assessor and client is to ensure that, between them, they have a clear and shared understanding of the required potential outcomes from the assessment and that the client is aware of any potential constraints under which the assessor may be operating. (ISO 14015 does not offer guidance on intrusive investigations, and more discussion will be undertaken on this in the concluding chapters of this publication.)

At this stage, it is probably sensible to reproduce the logic of the process, and the schematic in Figure 3 identifies this.

Some detail of the sort of risk assessment methodology that might be employed by the assessor in developing their evaluation, particularly of contaminated land, is described later in this book. It is a frequent complaint, especially from the legal fraternity, that these risk assessments are not always either clearly explained or their validity for the proposed assessment properly established. Again, one would have to consider the dissatisfaction identified in the KPMG survey and, in particular, the evidence of insufficient breadth of discovery, in order to recognize the potential for failure here. Great emphasis should be placed in the opening rounds of the transaction in ensuring clarity of the client's intention for the assessment as well as the assessor identifying constraints that may arise in attempting to obtain validated results.

ISO 14015 also reiterates (ISO 14015, section 2) a number of definitions common to other international standards, in particular ISO 14001 (a specification for an environmental management system), and include the following.

What is environmental assessment?

- Environmental aspect (2.6), *an element of an organization's activities, products or services that can interact with the environment.*
- Environmental impact (2.8), *any change to the environment, whether adverse or beneficial, wholly or partially resulting from an organization.*
- Organization (2.12), *company, corporation, firm, enterprise, authority or institution, or part or combination thereof, whether incorporated or not, public or private, that has its own functions and administration.*

Additional definitions, which may be helpful for the reader, include the following.

- Intrusive investigation (2.11), *sampling and testing using instruments and/or requiring physical interference.*
- Validation (2.15), *process whereby the assessor determines that the information gathered is accurate, reliable, sufficient and appropriate to meet the objectives of the assessment.*

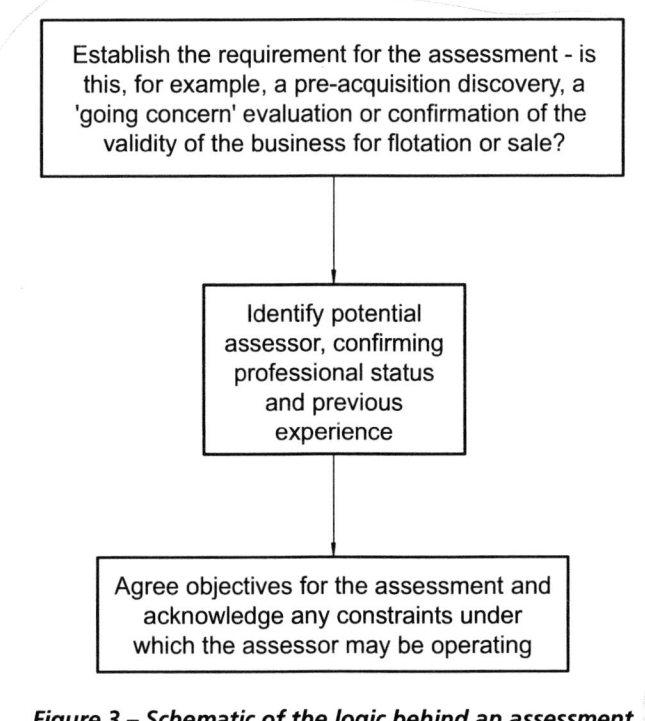

Figure 3 – Schematic of the logic behind an assessment

Some of these definitions may not be immediately familiar to the reader but it has been an ambition of ISO Technical Committee 207 (ISO/TC 207), the Environment Management Committee within the International Organization for Standardization (ISO) (www.iso.ch), to bring consistency in the use of its terminology across environmental standards and to provide ease of translatability from English into the other official languages of ISO – French, Spanish and Russian. In this context, the development of this standard and, because of the international authority that is brought by ISO documents, its potential use in many markets should bring a greater understanding of the terminology and processes.

Conclusion

The environmental assessment is a complex and comprehensive review of an organization's sites and operations, with a view to determining environmental aspects that may have an adverse or beneficial impact on an organization's financial well-being – the 'business consequences'. Its execution requires assessors with considerable environmental knowledge, commercial experience and intellectual skills, such that they can assist in contributing to the identification of business consequences. The assessment contributes to the overall environmental due diligence process, and may lead to recommendations on:

- intrusive investigation and the potential for remediation of contaminated land or the removal of potentially dangerous asbestos;
- the review of raw material and component sources and the use of recycled material;
- changes to suppliers with more sustainable working practices.

Chapter 2 – The client/assessor relationship

Introduction

The potential complexity of this relationship, with the client often being a 'proxy', has been referred to earlier in this book. With clients who may be national governments, multi-national or trans-global businesses or simply a growing commercial enterprise, the scope and scale of the requirements for environmental due diligence and the expectations from an environmental assessment, the potential scope for the assessor's consideration may be considerable. Indeed, it is patently obvious that, at the larger end of the scale, the environmental assessment and the overall environmental due diligence result may only be achieved by the use of a well-briefed team with, among others, sector-specific expertise in economics and the commercial aspects of the business as well as a wide range of environmental skills including soil engineering, geology/hydrogeology and biology.

Their access to sophisticated proprietorial software for risk assessment relating to air and waterborne pollution and for soil remediation assessments and to the subsequent engineering solutions is essential. However, in completing the fundamental elements of the environmental assessment, less initial sophistication is necessary. Initially, the client has some fundamental responsibilities that extend beyond the initial determination of the need for the assessment. The standard suggests (ISO 14015, 3.1) that the client should define the objectives for the assessment and determine the scope and criteria for the assessment. These may be provisional considerations pending the recruitment or assignation of the assessor to the task. In almost every circumstance, reconfirmation of the scope and objectives should then be undertaken between the client and the assessor.

The client's role and responsibilities

There follows a recommendation in the standard (ISO 14015, 3.4f) that the client should define which parts of the assessment will be conducted by the assessor and which parts will be the responsibility of the client. Care is needed in the interpretation of this requirement as the assessment has a number of significant activities:

- planning;
- information gathering and validation;
- evaluation;
- reporting.

Given the often limited grasp of the activity by many industrialists and, one suspects, some of their legal advisers, the assessor's ability to influence the choices may be important and will, in any case, depend upon the nature of the assessment to be undertaken – overt or confidential, internal or external asset, local or remote, and so on.

Unless the client has the immediate involvement in an internal assessment, the planning of the assessment is probably best left in the hands of the assessor. The assessor (or assessment team) is likely to have reasonably broad experience of the process and may, at least in overall scope, be better able to schedule the assessment and identify key elements of the asset for investigation. This is, of course, a commercial activity – time is money and the client will want some control over the overall costs for the assessment and the assessor – team or individual – needs to work within reasonable guidelines on time and the scale of proposed or potential expenditure.

In regard to the process of information gathering and validation, several issues arise. For the client, especially if they also represent the assessee (the subject of the assessment), there is the question of objectivity and understanding. Are they sufficiently familiar with environmental compliance and the detailed objectives and targets for their environmental management system? Can they supply, on an objective basis, material for the assessor's evaluation? Do they know or have ready access to the external sources of information that will complement the in-house material? If, as is likely, the answer to these challenges is 'no' and there is an inability to provide suitably knowledgeable staff to assist, the balance of experience on the assessor's side should be allowed to prevail. The client, as assessee, should, at least, facilitate the quick and accurate provision of information.

However, the balance of authority swings more in favour of the client when the issue of evaluation arises. With the benefit of the assessor's validated findings and their anticipation of some of the costs of, for example, improved emissions management or remediation, the client will be best able to consider the potential impact upon their business – how their capital investment programme might need to be re-phased or prioritized; whether or not the continuation of a production line may be valid in the face of increased cost or reducing availability of raw materials; whether energy taxes are liable to render the business less viable.

These examples refer, of course, primarily to the internal assessment. What then of the assessment of remote assets, whether overtly or covertly? Again, the client is likely to be the best judge of many of the product- or market-related issues, and the assessor might need only to apprise the client of the issues of environmental significance that might need to be addressed, with a reasonable estimate of the market costs for effecting the changes.

Once the initial assignation of general responsibilities and areas of authority have been established, some additional issues (ISO 14015, 3.1g-l) need to be clarified, and include the following.

- *The identification of priority assessment areas, if appropriate.* Either client or assessor may, on the basis of previous experience or sector-specific knowledge, identify certain physical assets or market-related activities, for example, land and premises with an extended history of use in a variety of industrial activities or competitor technology, which will contribute to the understanding of business consequences potentially arising.

- *Contact with the representative of the assessee, if appropriate, to obtain full cooperation and to initiate the process.* Where the assessment is being undertaken overtly, this is a straightforward commercial courtesy, whether the assessee site or asset is an 'internal' one or belongs to a third party. If the assessment is being undertaken in support of a covert environmental due diligence, care needs to be taken in regard to a number of activities. Site surveillance for evidence of controlled or careless emission policies for solid and liquid waste, for trade effluent and emissions to air and for general management of environmental issues requires considerable discretion. The investigation of public records containing relevant information and for which some identification is required needs to be handled sensitively, perhaps with adequate justification for alternative uses for the information. Where direct questioning or interviews are not possible, discrete discussion in the neighbourhood may alert the assessor to the less obvious site history and current operating circumstances.

- *Approving the assessment plan.* Time may be a very critical issue, with the requirement for a completed assessment and validated evaluations presented in a short period of time. This may require the mobilization of a team of specialists active in the various areas of site surveillance, archive and market research, emission modelling, laboratory testing and so on. Access to these resources is as important for the assessor as the financial resources to commission such help. This leads to a need for the next list item.

- *The provision of the appropriate authority and resources to enable the assessment to be conducted.* This needs no real explanation but assessors engaged to undertake third party assessments need to be alert to the financial constraints and any commitments to commercial confidentiality that may, correspondingly, be required.

- *Providing the assessor with the information necessary to undertake the assessment.* In an overt assessment situation, the client, either directly or through the empowerment of subordinates, may facilitate this straightforwardly. If the due diligence is being undertaken covertly, then the assessor has a more difficult problem in identifying performance statistics. Potential sources of information are discussed in later chapters but, especially in overseas markets, information that might be accessible in the more carefully regulated markets in North America, Europe and some countries of the Asia Pacific Rim will be absent elsewhere. In many cases, records of relevance may have been destroyed in wars or unavailable due to administrative inefficiency or the absence of adequate regulatory activity.

- *Receiving the assessment results and determining their distribution.* This, in a normal, particularly internal assessment situation, is a straightforward administrative exercise. However, where the environmental due diligence is being conducted in circumstances of great confidentiality, much greater discretion may be required. It behoves the assessor to ensure that lines and methods of communication are clearly established and, if appropriate, the identification of persons with whom contact is to be exclusively maintained.

The representative of the assessee

In the conventional assessment, there is a role for the representative of the assessee defined (ISO 14015, 2.13) as 'the person authorized to represent the assessee'. It is to the representative of the assessee that responsibility is given for the provision of information about and access to the asset under examination. Theirs will be the role of informing relevant staff members about the assessment and its purpose, of helping with the organization of interviews, ensuring safety and that suitable assistance is made available to the assessor. Of course, this role is irrelevant if the assessment is a discrete one and cannot be provided if the site being assessed is an 'orphan' – abandoned, with no known ownership.

The assessor

The standard (ISO 14015, 3.3) identifies a significant difference between the role of the auditor and that of the assessor. It suggests that the auditor is largely dealing with the verification of reported information and drawing conclusions as to the effectiveness and suitability of, typically, an environmental management system. In the assessor's case, they are dealing frequently with the requirement to research information that is sometimes new, in a relatively short period of time, often in discrete circumstances and may be required to provide evaluations based on their professional opinion and experience rather than hard evidence.

The standard requires that the assessor uses diligence, knowledge, skill and judgement, maintaining discretion and confidentiality unless required by laws or regulation to do otherwise. Given this not inconsiderable responsibility, the assessor's responsibilities are quite onerous. Their relationship with the client and the need to agree objectives, scope and criteria for the assessment and a reporting methodology have already been discussed. They also have the responsibility for preparing the assessment plan for the client's approval and winning title to the authority and resources to complete the assessment. In addition to these, the assessor, or the team that they lead, are required to undertake other actions.

- *Creating and maintaining working documents, such as checklists and protocols.* The derivation of these will be more complicated for the assessment of third party assets, especially those under foreign jurisdiction. The consolidation of an understanding of relevant legislation and its application may be difficult, especially where the regulatory regime has a poor record of enforcement and there is, perhaps, no case precedent available for offences against the local environmental regulation. The assessor/assessment team may require a knowledge of foreign market charging regimes for water, sewerage and other services, such as waste disposal, of the availability of remediation contractors and the relevant technology for the treatment of contaminated land or the removal of asbestos and of the availability, at an economic rate, of 'best practice' environmental technology and the maintenance support for it. In a world where there is increasing attention being paid to corporate social responsibility, where the consequential issues of working practices, economic contribution to the local community, unsustainable exploitation of raw materials and corrupt administrative or market processes may arise, the assessor/assessment team may need to cast their search for information quite widely. Working documents should reflect these challenges.

- *Ensuring that the necessary skills are available to meet the assessment objectives and if they are appropriate.* The assessor/assessment team will need a fundamental knowledge of environmental technology and working practices that prevent pollution and contravention of environmental legislation. However, the team may need supplementing with skills such as biology, hydrogeology/geology, economics, soil engineering and engineers from construction, mechanical and similar disciplines. For some complex assets, further knowledge from independent industry specialists may be necessary.
- *Obtaining the client's approval of the assessment team.* The client may be less interested in a review of individual skills than an assurance that there is overall team competence. They may also be interested in the techniques that the team will use in developing some of their models and assessments to ensure that the evaluations are robust and reproducible.
- *Obtaining initial information.* This should be readily available for the internal or overt third party assessment – indeed, a call for documentation may provide much of the relevant initial information. Elsewhere, the Internet may provide essential market information, details of environmental regulations, news of the assessee's possible violations of regulatory compliance and general market standing. Satellite and other airborne surveillance photography is capable of providing quite detailed information about land and buildings which are the subject of assessments. A preliminary reconnaissance for covert assessments is, therefore, perhaps more readily available than was the case a few years ago.
- *Assigning members of the assessment team to conduct the component parts of the assessment.* Essentially a straightforward management function, this can become complex if the due diligence exercise is a pan-national or multi-site exercise. Logistics, control and communication become significant issues for the assessment team leader.
- *Gathering and validating information in accordance with the assessment plan.* This is the essence of the assessment process. Later chapters of this book will offer the reader some options here for sources of information. However, it has already been mentioned that it is the consolidation of that information and its interpretation that is important. A key role, particularly for a team assessment, will be the careful filing and documenting of information, its source and potential reliability and the extent to which it has been corroborated.
- *Determining business consequences if requested by the client.* We have already discussed the client's assumption of responsibility for determining business consequences. However, if the results from the assessment and the overall environmental due diligence point to intrusive investigation of the assets, the client is likely to depend more heavily for some guidance from the assessor as to how this and any subsequent remediation may be progressed, with its potential impact on plans for both capital and revenue expenditure. Chapter 6 is dedicated to this and to issues such as the soil conditions to which the site will be restored, the nature of any engineering works that might be necessary, their cost

and how cash flow might be managed to avoid undue pressure on the business function and, finally, where suitable indemnification might be obtained to insure against any future potential claims.
- *Preparing and providing the report to the client, if requested*. Again, commercial skills are required to compile a suitable report, perhaps with illustrations, worked examples of technical assessments and tables showing performance comparisons of relevance. Again, the nature of the assessment report will depend upon the overall context of the environmental due diligence. Sometimes, all that is called for is a verbal report.

Assessor qualifications

The Introduction identified the absence of formal assessor qualifications or a professional body that oversees assessor behaviour. However, the standard (ISO 14015, 3.3) identifies some criteria through which a more informed selection might be made. They include education (to a level unspecified), training (for qualifications unspecified) and relevant work experience. The ISO auditing standard ISO 19011 [2], has changed the emphasis away from auditor qualifications and more towards auditor competence, demonstrated personal attributes and demonstrated ability to apply knowledge and skills. In this respect, potential clients might wish to be satisfied by the curriculum vitae of the assessment team and the list of their previous clients.

The standard (ISO 14015, 3.3, final paragraph) specifies additional requirements for knowledge of and competence in:

- relevant laws and regulations and related documents;
- environmental science and technology;
- economics and the relevant business area (this refers more to sector-specific knowledge than the geographical zone);
- technical and environmental aspects of commercial operations;
- facility operations;
- assessment techniques.

The client will doubtless make a judgement on the choice of assessor depending upon, perhaps, references from previous clients. This may, of course, be a little difficult, given the sensitivity of some of the work that the assessor may have undertaken.

Conclusion

The client/assessor relationship is a particularly important one, especially if the delivery of the business consequences is to be undertaken by the assessor. The client needs to be able to place considerable faith in the assessor's integrity, environmental insight, technical skill and commercial *nous*. In the best of circumstances, the assessor will be working with a clear understanding of the client's requirements, will have revealed their assessment

The client/assessor relationship

methodology, technological and other sources of help available to them and be involved with the client in determining the business consequences.

In a working situation where the assessment is being undertaken with the assessee's compliance, the assessor should have the capacity to draw out a comprehensive understanding of the historic and current conditions prevailing at the site or within the organization.

There are no qualifications or assessments of competency for environmental assessors. The client is prudent to follow up references and ask for work examples, where these can be provided.

Chapter 3 – Assessment planning

Introduction

The assessment process involves its planning, the accumulation and validation of information and evidence, its evaluation and the compilation of the final report, whether written or verbal. Ideally undertaken in detailed discussion with the client or their representative, the process is designed to identify environmental aspects with potential business consequences. These are often thought of in a negative context, especially as a potential cost to the business. However, if the client is so minded, the assessor may also be instructed to identify any potential business opportunities that might be exploited as a result of their investigations.

In the current political and regulatory environment, funding derived from the Landfill Tax Credit Scheme, and capital allowances for investment in more efficient heat and power generation processes, may prove attractive. The acquisition of 'brownfield' sites may also elicit support from local government and other similar bodies.

Objectives and scope

In most assessments, the objectives will be client-defined and primarily relate to the identification of environmental aspects and the potential business consequences. In some cases, the assessment will be made of abandoned, or 'orphan', sites, for which no such guidance is available. In this case, the assessor will need to consider some, more general, potential options, especially if the assessor is being asked to consider a disposal or development strategy for the site. In this case, their preliminary planning might wish to take into account:

- local/regional economic development strategies;
- local infrastructure requirements for residential or industrial developments;
- the level of remediation required to render the site fit for use, if contamination is identified;
- the need for risk assessments in the event that historic pollution has occurred and has migrated to or is migrating from the site;
- potential purchasers and the need for indemnification.

Some larger sites, such as military bases (see Figure 4), that, with de-militarization proceeding apace in northern Europe, were becoming plentiful at the end of the twentieth century, provide an intriguing combination of challenges. For example, some of the issues that can arise as a result of a demilitarized base are contaminated land, utilities and roads infrastructure, underground facilities and hazardous (waste) materials on site.

Environmental Due Diligence

Figure 4 – An example of a demilitarized base

Whether or not the client has identified the need for the assessor's evaluations of business consequences, agreement will be needed on the scope for the assessment. The standard proposes (ISO 14015, 4.2.3) a number of possibilities for consideration, which include the following.

- *Categories of environmental aspects to be assessed.* The environmental aspects typically under consideration will include discharges to air, land and water, including the off-site disposal of waste; compliance issues, including, for example, packaging waste obligations, the release of volatile organics or persistent organic pollutants, greenhouse gas management, packaging and product-related issues, in regard to market acceptability and raw materials consumption. If the client is considering the potential acquisition of the site for other uses, some of these aspects may no longer be relevant. Certainly, in the case of the disposal of an industrial site for non-commercial use, it is largely only the issue of residual land contamination that might be considered.
- *Any environmental impacts that other sites or organizations might have on the assessee.* The assessor will need, potentially, to undertake some off-site reconnaissance to identify the characteristics of neighbouring businesses. These may, of course, be relatively harmless service-oriented enterprises. However, in the case of, particularly, production activities, a knowledge of the potential hazards/pollution arising and whether these are husbanded responsibly is necessary, together with an understanding of the potential routes for migration of any potentially site-threatening materials. The assessor should be able to identify the existence of any of these on the subject site.

Assessment planning

- *Physical boundaries of the assessee (e.g. site, part of site)*. The enormity of some sites, such as the military bases mentioned above, major docks such as those in Cardiff, major manufacturing works such as British Steel in South Wales and the Central Midlands and sites in the mining industry provide examples of extensive sites where some limits may be imposed on the assessor in their work. This may be, for example, to allow phased disposal of the sites and generate cash flow to fund remediation or the progressive development of new infrastructure.
- *Adjacent and nearby sites, if applicable*. In addition to the presence of other potentially polluting or possibly polluted sites, the assessor needs to be alert to the existence of such areas as Sites of Special Scientific Interest (SSSIs), areas of outstanding natural beauty (AONBs) and any habitats that may be liable for protection under the Habitats Directive. The presence of these may have an impact on the marketability of the site and the uses to which it might, potentially, be put. The policies affecting development in these areas are likely to be addressed in either local or regional government planning policies.
- *Organizational boundaries, including relationships with or activities involving contractors, suppliers, organizations (e.g. off-site waste disposal, individuals and former occupants)*. Those who are familiar with environmental management systems (EMSs) will instantly recognize the challenge that third parties make to the environmental integrity of an organization. In particular, the role of on-site contractors in regard to their consumption of energy and their general approach to waste disposal. The EMS asks the sponsor to identify the potential for pollution in normal and abnormal conditions. Anecdotal evidence suggests that plant shutdown and the deployment of third party maintainers on site is a prime example of an occasion when accidents and accidental pollution occur.

   The issue of past site occupants or employees is very valid. Current legislation (e.g. the Environmental Protection Act 1990) makes it possible for the cost of remediating historic pollution recoverable where the originator of the pollution can be identified. The pursuit of this cost is unlikely to be straightforward and will not, therefore, fall within the assessor's remit. However, discussion with former employees at the site, where these can be identified and their evidence corroborated, can be useful sources of information on site management and the occurrence of potentially polluting incidents.
- *Time period covered (e.g. past, present and/or future)*. This element of the scope is qualified further with regard to the activities of the assessee and/or the client (e.g. continuing present operations, plans for change, expansion, demolition, decommissioning, revamping and with regard to development of the criteria (see ISO 14015, 4.2.4). Here the standard lacks some clarity. The time element is influenced very much by the requirements identified in the second element of the qualification. The client is perfectly within their rights to set limits on the periods in which environmental aspects should be considered. The redevelopment of the site would, for example, possibly mitigate for the foreseeable future any requirement for a contaminated land investigation to be undertaken, unless the current regulatory climate required it. They might simply ask for an assessment to be undertaken to consider the viability of the plant as a 'going concern' and with current raw material and utility streams. On the other hand, they may be

seeking a comparative assessment, setting the current 'environmental footprint' against those of a new process. The ultimate assessment would entertain historic pollution, current environmental performance, especially in regard perhaps to compliance and the sustainability of the processes against a sale or potential flotation in order to identify the potential capital value of the asset.
- *Business consequences cost threshold, if applicable.* Perhaps a difficult concept to grasp, but here the client is simply saying: 'If the assessment reveals a potential financial impact on an investment decision beyond so many tens of thousands of pounds, suspend or abandon the assessment.'

Assessment criteria

The standard (ISO 14015, 4.2.4) attempts to identify potential criteria for the assessment, against which the information that has been gathered and evaluated will be assessed, and offers three examples.

- *Currently applicable and reasonably foreseeable legal requirements (e.g. consents, permits, environmental laws and regulatory policies).* The European Community has passed over 300 pieces of environmental legislation, and in extreme cases, chief executives or board members and functional managers with environmental responsibility render themselves liable to a maximum fine of £20 000 and/or 6 months in jail for serious breaches of regulations. In this case, compliance is of critical importance. This is, of course, in addition to any corporate liability that may arise and for which additional penalties may be incurred. What is equally important is to have a clear understanding of future legislation (with its potential cost implications). This is, to some extent, made easier by the scale of consultation at both European and national level – the soon-to-be-introduced Waste Electrical and Electronic Goods (WEE) Directive has been portended for five or more years, while industry groups stood their corner on controversial aspects of it. One other vital aspect of legalization is to understand the thresholds beyond which compliance becomes essential. Producer Responsibility legislation is distinctive and, at least in smaller companies, the scale of operations might be at least as profitable if, for example, activities are managed at levels lower than the thresholds for compliance.
- *Other client-defined environmental requirements (e.g. organizational policies and procedures, specific environmental conditions, management practices, systems and performance requirements, industry and professional codes of practice and conduct).* The standard has 'bundled' a number of topics together here in a manner that is less than clear as to the consideration. The reader may wish to infer that the potential purchaser is looking for behaviour within the assessee that makes an acquisition more synergistic. The alignment of polices and procedures in regard to the sustainable conduct of the company makes a consolidation of activities rather easier. If management practices and the supporting systems are intended to deliver environmental compliance and commercial sustainability, so much the better. The acquisition may, however, be one intended to provide additional skills, or manufacturing capacity hitherto unavailable to

Assessment planning

the potential purchaser. In this case, the sector codes of conduct and working practices may require some assimilation. If this is the case, does time permit and are the resources available for the assimilation of this additional understanding and discipline? In regard to specific environmental conditions, these might include access to, for example, landfill or other waste disposal facilities, cooling water, minerals and other raw materials.

- *Requirements, claims or potential claims of interested third parties (e.g. insurance companies or financial organizations)*. The concept of ethical investment has brought fresh interest to the requirement for sustainable management, the reduction and avoidance of pollution, ethical and sustainable sourcing and social responsibility. If the acquisition is intended to provide security for the raising of additional capital, or is to become part of the portfolio of a listed company, evidence of poor environmental performance or unsustainable behaviour may prejudice the availability of this capital, or cause the share price of the parent company to fall. In the case of insurance companies, the requirements for improved environmental performance as a condition of the issue of environmental impairment liability insurance (EIL) or, indeed, the lack of any because of conditions that make insurance impossible to obtain could be an influential consideration.

© Sinclair Knight Merz 2004

Figure 5 — Prescribed industries

- *Technological considerations*. This will be one of the most critical issues for a potential purchaser, especially of a company or site described as a prescribed process, where compliance with the Integrated Pollution Prevention Control Directive (IPPC) is required. A theme of IPPC is 'best practice environmental option', which commits organizations to the use of the best technologies for the prevention or reduction of pollution. The criteria 'not entailing excessive cost' was a feature of the UK's earlier Integrated Pollution Control (IPC) regulations, which allowed the permitting authority some discretion on how rapidly the 'best practice environmental option' was installed. This no longer applies, and the

Environmental Due Diligence

potential purchaser of such a prescribed process needs to be alert to the cost implications for upgrading pollution prevention (see Figure 5), monitoring and measurement equipment. However, IPPC goes further in seeking measures for conserving energy and reducing waste. Inevitably, these may well mean new investment in better facilities for production and for the production of heat and power.

Even with the appropriate equipment in place, the organization may not be able to guarantee a smooth passage. Non-governmental organizations (NGOs) espousing public health concerns abound – sometimes single-issue concerns, but many able to mobilize public opinion and postpone the moment when operations can commence by demanding public enquiries. These can be long and costly. The UK cement industry, with its dependence for economic operation on a variety of waste-derived fuels, has been a victim of such delays.

Assessment plan

At this point in the standard (ISO 14015, 4.2.5) the reader will identify guidance on the Assessment Plan. While many of the items on the list are straightforward, easily understood or expanded on elsewhere in this book, some of them are described in more detail:

- identification of the client, the representative of the assessee, and the assessor(s);
- the assessment objectives and scope;
- the assessment criteria;
- priority assessment areas. Acquisitions and mergers are often undertaken in circumstances where speed is of the essence. The decision to, in particular, purchase a business or site may depend, for example, upon marginal financial considerations that can be decided on the basis of a few critical details. The assessment may, therefore, be decided on one or two critical views, perhaps limited to process and pollution management. Such a decision would obviate the need for wider reconnaissance and investigation;
- roles and responsibilities;
- the working language of the assessment and associated reports. Where the acquisition or disposal is a feature of cross-border, international trading, the choice of language may have some critical connotations where the influencing of foreign stakeholders may be concerned;
- anticipated time and duration of the assessment;
- assessment schedule;
- resources requirements (e.g. human, financial, technological). The assessor should be able to demonstrate to the client that they have access to the appropriate skills and the financial resources to support a team in the field. International travel and accommodation may be required, together with the employment of local agents, car hire and access to local legal and technical help. It is worth reiterating that the personal skills should not only encompass environmental technology and legislation, but a

comprehension of the industry sector and the commercial pressures which exist in that sector;
- an outline of the assessment procedures to be used. The client can be given considerable comfort by the assessor if they are prepared to elaborate on the logic of their processes. Such logic may include a review of process flow information, an analysis of waste arising, investigation of permits and authorizations, site reconnaissance and observation, interviews with employees, with regulators and the general public. Where, subsequently, intrusive investigation becomes an issue, reference to the soil standards required from remediation, the investigative and remedial technologies to be used and the techniques for the assessment of risk to sensitive receptors should be discussed;
- a summary of the reference documents, working documents, checklists and protocols to be used. These could include industry codes of practice, interview *pro formas*, guidance notes from regulatory bodies, local government planning policy documents and other similar material;
- reporting requirements. It has already been suggested that some assessments are carried out in circumstances of extreme confidentiality. The creation and circulation of any paper or electronic report on these may be forbidden;
- confidentiality requirements. At worst, normal rules for commercial confidentiality should apply – fundamentally, an assessor should neither publish references to nor discuss with third parties the client's instructions, the nature of the assessee or their assessment findings. In addition to any normal commercial discretion, the assessor may, in any case, be bound by a confidentiality agreement.

The assessment plan should be presented to the client for formal approval, whether this presentation takes the form of a face-to-face interview or a written or other form of brief.

Summary

The conditions potentially existing in a site or organization can be varied and it is important for the assessor to have some indication of the objectives that the client has for it. Aside from any legislation affecting contamination or manufactured goods produced by the organization, business consequences may be influenced by tax concessions or incentives, access to raw materials and changing consumer preference.

Technological developments, driven either by competition or regulatory influence, may mean significant future investment. It is productive if the assessor has objectives and a plan that offer clear direction when they have made discoveries which exceed the client's capacity to address them. The assessor should be prepared to interrupt the assessment and discuss with the client its premature cessation.

Chapter 4 – The assessment process: information gathering

Introduction

The phase in which information is gathered and validation takes place is the phase in which, despite the prescription of the guidance in the standard and the potential for client direction, the assessor needs to demonstrate intuitiveness and professional understanding in order to deliver a result. In this section of the standard, the guidance given largely anticipates freedom of access to knowledge from a compliant assessee. Additional comment has, therefore, been provided on addressing the more clandestine assessment.

The important aim for this phase of the assessment is that information gathered is complementary to the scope and objectives agreed for the assessment and that it is sufficient, relevant and accurate for the derivation of the findings.

Information

The following sources or identity of potential information are highlighted in Practical Help Box 1 of ISO 14015, 4.3.1 (see Table 1).

Table 1 – Sources of information identified in ISO 14015, Practical Help Box 1
— Location
— Physical characteristics (e.g. hydrogeology)
— Adjacent sites
— Raw materials
— Land use
— Site sensitivity
— Materials storage & handling
— Emissions and discharges to air, water and soil
— Waste storage, handling & disposal
— Emergency preparedness
— Storm water
— Fire prevention & control
— Hazardous materials
— Spills
— Occupational and public health & safety
— Legal and organizational requirements
— Legal, organizational or other non-compliances and nonconformities
— Relationship with external parties

Environmental Due Diligence

The implications for each are discussed here below.

- *Location.* The subject site's relationship with the local community – industrial or residential – and its adjacency to sensitive habitats needs to be clearly described.
- *Physical characteristics (e.g. hydrogeology).* With a view to the potential for contaminated land, an understanding of the geology and hydrogeology of the site is important. If soil and water contamination are a possibility, knowledge of the routes for migration and the location of potentially sensitive receptors can be helpful. A complementary consideration would be the meteorology and topography of the surrounding area. If airborne pollution is an issue then, using predictive modelling, some assessment might be made of the general dispersion and potential grounding of plumes and the location of potential complainants. Reference to the local infrastructure might also be useful if development/redevelopment is a consideration – statutory nuisances such as noise, vibration, dust, odour and litter – can antagonize local communities if not controlled properly.
- *Assessee, adjacent and nearby sites.* In the context of the location, a description of the physical characteristics of the subject site and the neighbourhood is helpful. Any written description can be usefully supported by aerial photographs – vertical or oblique projection, maps and digital video recordings. A description of the existing land use should be cross-referenced to the local structure plan to ensure that re-allocation of use for the area is not predicted, or that development/redevelopment may not be restricted in any way. This is an issue especially where 'orphan' sites or land designated for 'brownfield' development is concerned. Any reconnaissance of a site scheduled for a confidential or adversarial acquisition bid will need to be undertaken discretely, if not clandestinely.
- *Raw materials, by-products and products (including hazardous materials).* The competent assessor with sector knowledge will have a comprehensive understanding of the materials and the processes employed and the products and by-products produced. They should also know whether there are hazardous materials employed directly or indirectly in the production process and the nature of the waste streams arising. If time permits, the management of all aspects of the production process should be observed to ensure that the storage and handling of, especially, hazardous materials conforms to expectations in terms of regulation and codes of practice. An important item of legislation with which organizations will be expected to comply is the Control of Substances Hazardous to Health (COSHH) Regulations 2002. Records of risk assessment undertaken by the organization to comply with these regulations should be available.

Where controlled and hazardous waste are concerned, confirmation that a suitable audit trail is available, confirming the consignment of such materials to authorized disposers who have issued, in the case of the hazardous waste, the appropriate destruction certificates is necessary. Compliance here is regulated under the Control of Pollution (Special Waste) Regulations 1996. Additionally, observance of waste handling may also give clues as to the opportunities for recovery for recycling with attendant potential savings in waste handling and disposal.

The assessment process: information gathering

In the case of the storage of hazardous substances, the Planning (Hazardous Substances) Act 1990 and the threshold beyond which its observance is required can be important.
- *Materials storage and handling*. This is an often-ignored area of activity, where product loss control and energy conservation is poor. Receipt of damaged components and the mishandling of raw and other materials in transit can be both a source of lost revenue to the company and, in extreme cases, a source of pollution. Pollution remains a concern to regulators and insurance companies, due to the lack of control of liquid fuels and other chemical storage. The absence of bunding for tanks or security for product movement continues to promote spillages with the attendant costs of lost product and the clean up of any pollution.

  The management of warehouse lighting and heating is also often poor, with lighting left at high levels even when the building is unoccupied and heat escaping through uncurtained or unshuttered doors. These are often pointers to a company run without consideration for the environment or its longer-term sustainability.

- *Emissions and discharges to air, water and soil*. The assessor will be concerned to see any relevant permits and authorizations and to ensure that effluent consents, emission limit values (ELVs) and other thresholds are not exceeded. Any cases of non-compliance, which might prejudice the longer-term availability of authorizations and permits, should be established. If time and resources permit, external consultation with regulatory bodies, local government authorities and water treatment companies may be appropriate to establish any potential additional liabilities. In some exceptional cases, organizations operate their own land filling operations. The assessor will need to ensure that any such facility is appropriately licensed and that monitoring for any groundwater contamination is properly in place and, if appropriate, gas monitoring or gas capture facilities are functioning properly.

  The identification of historic events – spillages to water and especially to soil – may be more difficult to establish. Investigations here may extend to including the examination of operational logs, incident books, interviews with existing and retired staff, emergency services and, possibly, residents from the local community. A site examination may identify, through, for example, dead or distressed vegetation, soil discoloration and contaminated sewers and storm water drains, the potential scale of soil and other contamination. This element of the assessor's work is, nonetheless, fundamental to the identification of a need for intrusive investigation.

- *Waste storage, handling and disposal*. The costs of waste handling and storage often represent an area of unaddressed or properly managed opportunity for resource efficiency. Evidence of waste segregation and a commitment to recycling may indicate a company in control of its overheads and committed, perhaps, to sustainable methods of operation. Organizations also need to observe careful storage of waste to avoid littering and the creation of dust and odours – all statutory nuisances for which penalties, typically fines, can arise.

There are also legal issues associated with the handling of waste – compliance with the Environmental Protection Act 1990, Section 34 (Duty of Care) and the Control of Pollution (Special Waste) Regulations 1980. Controlled waste should be properly consigned with a licensed carrier and special waste, with its pre-notification procedures, requires the maintenance of an audit trail, demonstrating responsible disposal and a destruction certificate or, if appropriate, a recycling note from a licensed recycler. A competent assessor will ensure that these records are consistently maintained.

- *Fire prevention and control, spill containment and other emergency planning.* Especially in sites where hazardous processes are undertaken or where hazardous materials are stored, the assessor will expect to find a comprehensive emergency plan, dealing with simple spill containment or management right through to a full scale emergency plan, identifying plant closure and evacuation procedures. Indeed, many larger organizations with potentially hazardous activities on site are required to observe the conditions of the Control of Major Accident Hazard Regulations 1999 (COMAH) and the assessor will look to ensure that this in place if the criteria require it. However, it is not sufficient merely for the emergency plans to exist. The assessor will expect to see evidence of the plans being reviewed for relevance and also being tested or practised in whole or in part. Such practice may involve a desktop exercise, the exercise of an element of the plan and even, periodically or as determined by legislation, a full scale practice.

  There should be comments arising from any incident recorded in an appropriate document – incident or operational log – and, in the case of a major incident, some evidence of a formal enquiry into the events and of the incorporation of any improvements recommended. The plans should include recognition for the potential for environmental damage both from materials released during the incident as well as that arising from the use of chemicals and other materials used to treat the release.

- *Storm and floodwaters.* The assessor will be seeking to ensure that effluent treatment plants and storm water drain capacity are adequate to manage excessive and violent rainfall. This will include ensuring the integrity of the systems designed to keep potentially contaminated process water out of storm water drains and that the treatment plant has sufficient excess capacity to permit the treatment of water from the oily water sewers, etc. Drainage and sewerage are often colour-coded to ensure that containment materials can be deployed quickly where clean/storm water drainage is at risk.

  Floodwater offers a different challenge and should concern the assessor, given the increasing tendency for development to be undertaken in flood plains. Where flooding or tidal movement can threaten, some form of dam or barrier in underground sewers is desirable. Importantly, where floodwater can otherwise threaten a site, the assessor should ensure that tank farm bunding and containment for hazardous and other potentially polluting materials is installed.

- *Occupational and public health and safety.* Where time permits and the assessor is able to witness operations, there are several issues on which they should take a view. These could include, but are not limited to, the problems of noise, dust and vibration. Although personal protective equipment (PPE) can address some of these problems, the sustainable solutions include the provision of acoustic hoods on noisy equipment, the fitment of vibration dampers on rotating and similar equipment, the installation of noise and vibration suppressing floor coverings and of dust capture equipment on cutting and other equipment where dust occurs.

One issue that also demands the assessor's attention is in regard to asbestos. This has been used in a variety of applications including pipe and furnace insulation materials, asbestos and cement roofing, floor tiles, backing on vinyl sheet flooring, soundproofing or decorative material, fireproof gloves and stove-top pads, vehicle brake pads and linings, clutch facings and gaskets. Intrusive investigation may be necessary in order to ascertain the extent to which asbestos may be present in buildings, machinery and power generation equipment and in site and other waste present. If the fabric is in good order and air quality tests reveal no significant hazard from airborne fibres, it may be appropriate for the material to be left *in situ*. Guidance is given under the Control of Asbestos at Work Regulations 2002.

Although not of immediate concern here, there is the wider issue of the 'sensitive receptor' in relation to emissions and

- *Legal, organizational and other requirements, non-compliances and non-conformances.* Some references have already been made to essential items of regulation. The assessor will, in circumstances where the assessee is a willing subject, have access to the organization's Register of Significant Legislation if it has an environmental management system in place. Failing this, they will need to bring a knowledge and experience of applicable legislation to the assessment and ensure that the organization knows of, and is compliant with, regulations affecting, for example, discharges to air, water and soil, packaging regulations, handling of hazardous materials and so on. The directions given in the standard also recommend, where the assessor is given access to internal audit information, that a review of the non-compliances and non-conformances be undertaken. These, and the management review outcomes, give a clear steer as to the company's or organization's commitment to continual improvement in its environmental management performance and also to any underlying inconsistencies in that performance – communication, training, equipment obsolescence and so on.
- *Relationship with external parties.* Implicit in this reference are some, if not all, of the 'environmental stakeholders'. The relationship with regulatory bodies has already been discussed in respect to authorizations, permits and consents. Others with whom consultation may be appropriate include the following.
  – Neighbours, especially residential. An organization's 'licence to operate' in regard to permits, planning applications and other regulatory activities can be influenced dramatically by the public perception of its operations. Incident avoidance, regular communications with the community and, indeed, some positive participation by employees in community projects may help with perceptions of the organization's attitude to the environment and sustainability.
  – Insurance companies. Environmental Impairment Liability insurance is beginning to supplement General and Public Liability Insurance, from which the cover for 'sudden

and accidental pollution' is increasingly being withdrawn. Poor environmental performance, with a history of polluting events, may attract high premiums for this insurance and vice versa.
- Banks. Banks are now taking environmental risk seriously and an element of their financing charges is likely to reflect the environmental probity of the organization. Interruptions to or prohibitions on production can have a dramatic influence on cash flow and the consequent ability to service loan and overdraft repayments.
- Clients. Where the assessment is taking place in cooperative circumstances, discussions with clients and customers may reveal plans for changes in their requirements for services and components. The concept of the 'green supply chain' may demand improved processes, the introduction of a stream of recycled materials and the qualification of new materials. The assessor's discoveries here go right to the heart of the commercial implications for the organization's future and, indeed, the business consequences for any failure to adapt to the trend in environmental sustainability.
- Suppliers. The upward trend in Summer 2004 in the price of crude oil reflected not just the political tension in the major supplying regions but also the burgeoning growth of demand for materials in the Asia Pacific Rim, especially China. While not an immediate environmental impact, the access to raw material and components at competitive rates may become increasingly unsustainable without serious attention to energy reduction and conservation measures and improved product and process design (design for environment) for easier end-of-life recovery for potentially recyclable materials. This is also a legislative issue – producer responsibility legislation now influences the manufacture of packaging, vehicles and electrical and electronic goods.
- Environmental non-governmental organizations (NGOs). News from NGOs is not always good for the organization, but they are, nonetheless, passionate commentators on sector-specific environmental performance issues, corporate social responsibility and the issue of biodiversity. The NGOs, including bodies such as World Wildlife Trust, Greenpeace, Friends of the Earth, local wildlife trusts and other conservation bodies have demonstrated extremes of behaviour. These range from the development of more environmentally friendly techniques to address some problems of environmental pollution to the extreme situation where premises are occupied or entry is barred by way of protest at environmental pollution or unethical behaviour. The speed with which news now travels, especially over the internet, means that multi-national organizations must ensure that their activities can withstand robust examination for breaches of ethical and environmentally sound behaviour in whichever part of the world they operate.
- Investment analysts. We have not discussed the relationships of companies quoted on the Stock Exchange with investment analysts. This dialogue forms part of a routine exchange of information on (financial) market perceptions of the company's

performance, including its environmental profile and perceived environmental efficiency and discrete knowledge of the organization's internal thinking and, for example, the issue of senior appointments within the company. It may also be a dialogue generated as a prelude to public pronouncement relating to mergers, acquisitions or other financial restructuring. If information on such exchanges can be made available, albeit discretely, it may assist the assessor in forming their opinions. It is, of course, possible that information on the company has already been established by the Ethical Investment Research Service (EIRIS – www.eiris.org) or the Social Investment Forum (SIF – www.uksif.org or www.socialinvest.org), and which might be obtained by the assessor. Some of the information obtained may be speculative or an interpretation by the analyst of events predicted or which have occurred and so will require careful analysis by the assessor.

The standard next discusses (ISO 14015, 4.3.2) the need to examine existing documents and records to obtain a sufficient understanding of the site and/or organization, without unnecessarily duplicating prior investigative efforts. A practical help box (see Table 2) has been inserted into the standard to identify typical commercial documentations giving various insights into the organization's environmental performance and the management structure that supports it, or not, as the case may be. The standard encourages the assessor to identify documents, which will deliver corroboration of information, without needlessly duplicating work. Wherever possible, the assessee is encouraged to assist in the identification and delivery to the assessor of this material. A failure to assist in this way may limit the value of the assessment or delay its production, with the consequent potential penalty in terms of cost.

The assessor should, in the course of a normal assessment, maintain a data log, identifying the type, source, quality and reliability of the information to enable more effective validations.

Table 2 lists documents and their potential sources. It is neither an exhaustive list nor one that the assessor is obliged to address in its entirety. Depending upon the objectives and the assessment criteria, the assessment result may well be obtained without reference to this exhaustive list of possibilities. However, where discovery is proving difficult, it does offer some alternative sources that might be useful, whether the assessment is an overt one or a clandestine one.

The assessor is also encouraged (ISO 14015, 4.3.3) to observe activities within an organization and physical conditions on site to derive an understanding of current activities and past operations. In undertaking these observations, the assessor is encouraged to limit their discoveries to those derived from the natural senses – typically, vision, smell and hearing. It is, of course, a very practical matter now to take digital photographs with which to remind the assessor of issues and with which to illustrate any subsequent written submission to support the assessment findings. Table 3 gives lists of activities and physical conditions that the assessor may wish to see or observe.

Environmental Due Diligence

Table 2 – Example of documents and sources that may be considered in an environmental assessment of sites and organizations (EASO), as identified in ISO 14015, Practical Help Box 2

Documents	Sources
— Maps, plans and photographs — Historical records — Geological/hydrogeological records — Geotechnical records — Consignment notes/manifests — Safety data sheets (material safety data sheets) — Work orders — Monitoring procedures and results — Process documents (e.g. material balance) — Inventories — Containment plans — Other response plans — Health, Safety and Environmental (HSE) training records — Accident records — Permits/licenses — Organization charts (tasks and responsibilities) — Audits and other reports — Non-compliance and non-conformance records — Complaints — Company policies, plans and management systems — Insurance requirements	**External** — Government agencies (national, local, regulatory, planning) — Archives — Utilities — Commercial publications — Industrial codes of practice — Emergency services — Insurance bureaux **Internal** — Environmental, health and safety department — Engineering department — Asset management — Facilities management — Training department — Legal department — Finance and accounting department — Public relations department

The justification for such investigation has already been addressed in many of the preceding paragraphs. It would be simple enough to say that the assessor should, having referred to operating procedures, industrial protocols and specific contract requirements, be able to identity from their observations the degree of compliance with regulations and other requirements and the cultural observance of the organization's corporate requirements. This will enable the assessor to draw some conclusions as to whether there is substance in their records and general performance or whether it is, in fact, camouflage for a less than rigorous operation.

Table 3 – Examples of lists of activities and physical conditions to be observed

Activities

— Waste management
— Materials and product handling
— Process operations
— Wastewater management
— Property use

Physical conditions

— Heating and cooling systems
— Piping and venting
— Containment, drains and sumps
— Storage containers/tanks
— Utilities supply
— Stains
— Noise, light vibration or heat
— Odour, dust, smoke, particulate
— Surface waters and site landscape
— Site surroundings and adjacent properties
— Soil and groundwater conditions
— Stained or discoloured surfaces
— Damaged vegetation
— Landfills
— Plant equipment
— Material storage
— Hazardous materials, products and substances
— Discharges to water

Conclusion

The accumulation of information can be complex and painstaking, involving on-site observation, interviews and the consultation of records both on and off site. Aside from the obvious signs of pollution, poor operational practice and recorded incidents of environmental pollution, the assessor will be required to use intuition and experience to pursue all the information that they need. The speed and scale of international communication, especially that involving the Internet, mean that there are few secrets in relation to the performance of multi-national organizations. The use of the 'whistle-blower' as a valid means of obtaining first-hand information on unethical or illegal behaviour is increasingly acceptable. The existence of external environmental performance rating

Environmental Due Diligence

processes (e.g. EIRIS, Dow Jones, FTSE) may help, if not practically, in certainly screening a company's performance.

However, it is important that the assessor is methodical and carefully records activities and results before attempting any conclusions.

Chapter 5 – The assessment process: interviewing and validation

Introduction

The purpose of interviewing is to obtain information or to corroborate or augment information accumulated through observation of activities or the examination of documents.

Validated information is important as it provides the foundation for the evaluation process.

Interviewing

The technique for interviewing requires both carefully prepared questions and good interpersonal skills. The idea of carefully prepared questions is to ensure, as far as possible, that they are consistently asked of a number of interviewees so as to elicit as frank, objective and as informative answers as possible. Good interpersonal skills should similarly ensure that the interview can be developed beyond the 'stock questions' to ensure that the topic is fully explored.

The range of interviewees will depend, once again, on the nature of the assessment – a fully-supported assessment with ready access to the assessee and relevant personnel, or a discrete assessment, where the interviewing process may, of necessity, be conducted on a more casual basis. More care will be required in the latter circumstances to ensure that any information obtained is suitably corroborated.

The standard (ISO 14015, 4.3.4.2) provides a practical help box (no. 4) in which a number of groups of potential interviewees are identified. The obvious groups include:

- management with environmental responsibilities and those without;
- environmental specialists in the organization or those advising it;
- personnel responsible for on-site activities, including equipment operators;
- neighbours;
- internal health and safety staff.

Less obvious groups include the following groups of potential interviewees.

- Maintenance operators. When considering their environmental aspects, organizations are asked to consider normal and abnormal activities. Maintenance staff are those most commonly in attendance when the plant is being started up, closed down or under repair or maintenance. Their views are important in establishing non-routine activities that may represent a threat to an organization's environmental integrity.

- Former or retired employees. Of the two, it is likely that the latter group may be more reliable in providing historic information on an organization's operations and any incidents that might not be recorded elsewhere. However, departed employees may, in an era where 'whistle-blowing' is an accepted phenomenon, also provide discrete but revealing insights into an organization's working practice.
- Environmental regulatory agencies. Most of the leading economies have now enacted a comprehensive suite of environmental regulations. However, it is equally true that these are not supported by an aggressive enforcement policy and many of the regulatory agencies are consequently weak or almost invisible. Interviews with the officers of the regulatory agency may provide information on the degree of rigour with which enforcement is undertaken, with the consequent potential for organizations to 'flirt' with compliance.
- Fire authorities. In any site undertaking any substantial manufacturing or similar process, involvement with the local fire brigade is an essential part of the emergency planning. The local authority will have a perspective on the effectiveness of the equipment levels and staff training, on vulnerable locations in the vicinity of the site and, possibly, records of incidents at the site.
- Other emergency services, including health, and municipal services. Most local authorities have contingency plans for dealing with major incidents, evacuations and disasters. Their relationship with the assessee may be dictated either by the potential for problems on site, or the extent to which the organization can contribute suitably trained staff and equipment in support of these organizations.
- Legal advisers. These may have been involved in both an advisory role as well as supplying legal services where breaches in compliance have occurred and may have an objective view on some of the softer issues in the organization, including the corporate culture, the quality of management and internal communications.
- Contractors. It is becoming increasingly evident that those who supply materials and services are an increasingly important part of the efforts to avoid polluting incidents and breaches of regulation in regard to, for example, waste management and some statutory offences – noise, odour, litter, dust and vibration. Communication with them may reveal shortcomings in the assessee organization's procurement, safety and other protocols.
- Procurement personnel. Much depends on the procurement team for the acquisition of sustainable raw materials, the avoidance of hazardous materials, the procurement of services with suitable environmental awareness and, for example, the negotiation of waste disposal contracts. However, they are only as good as their training permits – they may be unaware of 'design for environment' principles and the problems of end-of-life disposal, although the concept of 'whole life costing' may be more familiar to them. However, discussion with this part of the organization may be beneficial in revealing this depth of understanding and, perhaps more importantly, the attitudes of major suppliers to the growing challenges of environmental probity.
- Former occupants. The ability to identify former site operators is important when assessing the potential for hidden pollution and, indeed, identifying the scale and success of any past pollution remediation and the existence, perhaps, of indemnifications on this work.

The list is not exhaustive but, nonetheless, provides the assessor with target groups from whom to derive relevant information. In approaching them, the scope of the questions is likely to invite (ISO 14015, 4.3.4.3):

- descriptions of employees' work and the way in which it is carried out now and was in the past; and
- information on site uses, conditions and history, with particular reference to events that have had, are having, or may have an environmental impact.

The standard (ISO 14015, 4.3.4.4) suggests that there may be limitations on the interviewing process and the information it might glean. Interviewees are, of course, under no obligation to participate or may be limited in their response by a lack of familiarity with the topics being discussed. Another limitation in the interview may be that of linguistic skills or unfamiliarity with technical language.

It is important that the assessor summarizes the outcome of the interview and identifies any conclusions they may have arrived at to the interviewee to ensure that the correct understanding or perception has been obtained.

Information validation

The standard (ISO 14015, 4.3.5) identifies several characteristics expected of the information accumulated – accuracy, reliability, sufficiency and appropriateness for the purposes of meeting the assessment objectives. If the information is compromised by failing to meet these tests, the assessor is directed to advise the client as quickly as possible. This is, of course, when ideal circumstances prevail – unfortunately, not necessarily the situation when the assessment is being undertaken in a remote (from the client) situation. Therefore, except in the most extraordinary circumstances, it is likely that any limitations will be advised to the client at a more convenient moment, if not when the final report is submitted. It is a characteristic of the assessment that the assessor is given considerable discretion in being able to exercise professional judgement. This discretion should not divert the assessor from reasonable efforts to derive corroborated and validated information.

Conclusion

The techniques for interviewing are well understood and regularly practised. However, in order to maintain consistency, questions need to be recorded and asked consistently of interviewees, where corroboration is important. The assessor should have sufficient knowledge of the process under assessment such that, where it is appropriate to break away from the prepared questions, they have sufficient knowledge to ask pertinent questions.

The identification of environmental issues should be straightforward, except where intrusive investigation is required – the identification of the below-ground pollution, its nature and concentration, potential pathways for migration and the potential risk to receptors may be

Environmental Due Diligence

a significant undertaking. If this is required, it will be a detailed part of the evaluation of the business consequences – not only requiring accurate assessment of the costs of the work, but also the implications for land values and use, once remediation or containment is completed.

Chapter 6 – The assessment process: evaluation of issues and determination of business consequences

Introduction

The evaluation of issues and the determination of business consequences may be undertaken in a variety of circumstances. In discrete circumstances, where an acquisition may be being considered, it is likely to be by the assessor alone with, perhaps, some dialogue with the client. In more overt circumstances, an assessment team including specialists may be assembled to present conclusions to the client. Ultimately, the best results are likely to be obtained by the assessor or assessment team working with the client.

Evaluation

The evaluation process described in the standard (ISO 14015, 4.4.1) promotes two stages in the evaluation process – the identification of the environmental issues and, then, the determination of the business consequences. In the more complex assessments, the assessor, as the primary observer and enquirer, will probably require a team that offers additional skills in technical subjects, finance and the law.

In chapter 3, we identified the fact that there were likely to be certain criteria associated with the evaluation stage of the assessment. These criteria could, for example, relate to the assessee's degree of compliance with legislation and its current status in regard to forthcoming regulation; to the security of supply of relevant raw materials and their potential for replacement with recycled materials; the sustainability of existing processes and so on. It is against such criteria that the evaluations will be undertaken, seeking to identify significant deviations from the criteria that may suggest liabilities or, indeed, benefits to the organization. There may be implications – increasingly so where issues of sustainable sourcing, community interaction and economic dependence are concerned – for the organization's public image. Most importantly, there are implications for costs, not so much for purchase of the asset as for the remedial or improvement work that might be required once in possession. If the assessment is being undertaken prior to a sale and contaminated land is discovered, a decision needs to be taken on whether the asset should be sold at a discounted price, reflecting its brownfield characteristics, or remediated with the provision on an indemnity. This latter course of action has been adopted by a number of petroleum companies selling assets for, especially, residential developments in urban areas.

Determination of business consequences

There has been, in recent years, a preoccupation with contaminated land and the not insubstantial costs associated with remediation. However, costs may arise in other areas of more current significance, such as the following.

- Modifications (future liabilities) to comply with the requirements for the Integrated Pollution Prevention Control Directive (IPPC). The objective within IPPC is for the deployment of equipment and processes representing the 'best practice environmental option' (BPEO) for environmental protection and pollution prevention. This precludes the continuation of end-of-pipe solutions and the deployment of more efficient processes consuming less energy and other raw materials; of effluent capture and recycling or pre-discharge treatment; of new processes to address the requirements of new 'design for environment' principles and their attention to whole life impacts; of waste management processes dedicated to optimized recovery for re-use, recycling or safe disposal and, of course, the training essential to keep up with these changes.
- Market requirements (current, potential or future liabilities). Consumer awareness is becoming more sophisticated with, for example, the emergence of the 'Fairtrade' brand supporting ethical and sustainable production. The desire for organically produced vegetables and fruits is having a significant impact on agricultural production. The emergence of a variety of non-governmental organizations is influencing manufacturing in newly developing countries and economies in transition, demanding reductions in the employment of child labour, improved occupational health management and other changes of significance to the transglobal business.
- Supply chain requirements (current, potential or future liabilities). Effective environmental management can significantly improve the opportunities for a term contract, combining quality with 'just in time' delivery of components designed for re-use or recycling and contributing to the easier dismantlement of products at their life's end. Such production requirements may require investment in research and development, staff training and the deployment of different production methodologies. The introduction of Producer Responsibility legislation has been instrumental in significant changes in the manufacturing and use of packaging, and the manufacturing and final disposal of vehicles and electronic and electrical goods.

In addition to the above, the standard (ISO 14015, 4.4.3) confirms the requirement to be able to put a price on the investment in pollution prevention technology, in new processes and training, in research and development and the general costs of meeting compliance.

The reader will now be able to understand why, at this stage in the process, the environmental assessor's capacities may be stretched to provide informed opinion on non-environmental issues and, probably, some of the technology that might be applied. The purchasing (or disposal) decision may hinge substantially on, for example, the informed view that they might bring to the investigation and treatment of contaminated land. However, if they have not been briefed on the client's strategic objectives nor, indeed, the potential value to them of any purchase and the congruency it may bring to their existing portfolio, then the assessor will find it difficult to provide further services.

Conclusion

The determination of business consequences, if left to the assessor, will be made against a set of criteria agreed at the commencement of the assessment. The client should expect transparency in the assessor's methodology when explaining their conclusions.

The assessor's ability, having recognized the environmental issues during the assessment, will be required to exercise commercial judgement and foresight when determining the business consequences, recognizing potential costs for the acquisition of relevant technology and training, predicting potential changes in markets and the timing and impact of future legislation.

Chapter 7 – The assessment process: reporting to the client

Introduction

In extreme cases, where confidentiality is at a premium, the client may require only a discrete verbal report. However, in the normal course, a written (paper or electronic) report would be expected.

Report contents

The report is primarily the assessor's responsibility (ISO 14015, 5.1) and the recommendation is that the format should be such as to enable the client to focus on the significance of the findings. The standard advises that the assessor should distinguish fact from opinion, to clearly identify the basis for the findings and indicate the relative uncertainty associated with any finding.

The standard first lists some basic information required for insertion:

- the identification of the sites and/or organizations assessed;
- the name(s) of the assessor(s) and the author of the report;
- the assessment objectives, scope and criteria;
- the dates and duration of the assessment;
- any limitations of the available information and its consequences on the assessment;
- any limitations, exclusions, amendments and deviations from the agreed scope of the assessment; and
- a summary of the information collected during the assessment and the results of the assessment.

Further information is discretionary, but the report can also include the following.

- The name of the client.
- The name of the assessee's representative.
- The composition of the assessment.
- The assessment schedule. The standard is unforthcoming on this element, but it suggests that the assessment should conform to some form of timetable of meetings (if the team is large enough to warrant it) to corroborate information and agree results and record formally the outside contacts made.
- A summary of the assessment procedures used. This might, in the simplest of assessments, identify no more than a desktop study with some benchmarking on a sector-by-sector

basis. However, the summary is likely to address the scale of the observations undertaken within the organizations, any reconnaissance undertaken around the site and its neighbours, the extent to which documentary, photographic and other evidence may have been examined and the scale of any interview programmes undertaken.
- A summary of the reference documents, checklists and protocols and other working documents used. The list of possibilities here is considerable, ranging from regulations and regulatory guidance on them, permits and approvals, industry codes of practice, management system documentation, operational logs, questionnaires for interviews, checklists, historic documentation and so on. Where the assessee is not cooperative, or the assessment is being conducted clandestinely, the scope of documents available for reference may be restricted and some dependence may need to be placed on library documents and other, perhaps historic, material in the public domain.
- Evaluation methods, and the basis upon which evaluations were made. The possible options for evaluation methods are quite considerable. They will relate to, among others:
  - compliance and the assessment of potential liability (fines and imprisonment) arising from cases of non-compliance. This would be risk-based assessment, considering evidence of past non-compliance within the organization, of examples of non-compliance by sector, and the precedents set;
  - environmental improvement. The costs of achieving the 'best practice environmental option' need to be benchmarked against competitor and sector achievements, to have the potential costs discounted for comparisons with other investment options and, perhaps, to consider the competitive advantage offered by being 'best/leader in class'. Many companies will have developed software methodology for factoring in the various options to identify the possible outcomes;
  - organizational performance. If an environmental management system is in place, audit records should provide evidence of its success in terms of application, training and continual improvement. Failing the availability of an EMS, the assessor's own observations and experience will be brought to bear;
  - product-based evaluations. These will consider raw material and component inputs, security of supply and options for the use of recyclable materials, legislation affecting end-of-life disposal and whole-life issues.

  There are likely to be other evaluation methods that could be considered and the report writer will need to consider the relevance of details. Many of the methods are likely to be qualitative rather than quantitative. Where intrusive investigation is considered or recommended, the report writer may wish to refer to some of the methodology used. This would, at this stage, be both speculative and discretionary.
- Results of the evaluation if conducted by the assessor.
- Recommendations regarding the next possible steps. Along with the results of the evaluation, this element of the report is perhaps the most telling and potentially useful. Effectively, it is the directional guidance the client needs to, perhaps, plot their negotiating stance, how they might exploit the organizations on acquisition and so on. Balance and conservatism is required in the recommendations.

The assessment process: reporting to the client

- Confidentiality requirements. These will be a reiteration of the classification and handling requirements agreed by the client at the outset of the assessment.
- Conclusions. The tenor of these remarks may be predetermined by the initial discussion with the client. Should they be framed in unconditional terms – effectively, a buy/do not buy conclusion. Or, might the assessor be required to give some conditional remarks and make suggestions for negotiating positions or other challenges that the buyer might make to win concessions.

The concluding paragraphs to the section encourage the provision of documentation to support the essential conclusions of the evaluation, especially if the work is to be revisited in the future, or if it represents an in-house case study for the application of, for example, future investment criteria or for a disposal/exit strategy.

The standard concludes (ISO 14015, 5.2/5.3) with the identification of the options that might exist for the report format and direction as to the ownership and possible circulation of the report.

Conclusions

The first critical decision the assessor will make is whether their information requires verbal communication or whether the client prefers a report in written or electronic format. Confidentiality and other terms of reference for the client/assessor relationship will be the deciding factor.

A written report will contain fundamental details about the assessment team, the scope of the assessment, relevant criteria and the identification of the environmental issues and the resulting business consequences. More detailed information can be provided by agreement.

Chapter 8 – Intrusive investigation

Whilst ISO 14015 excludes intrusive investigation, in practice there may be the need for this to be part of an environmental due diligence exercise.

Where ISO 14015 is used in the context of the sale and/or the re-development of land it will almost inevitably form part of a process that will require intrusive ground investigation. If the due diligence is during a merger or acquisition where redevelopment is not planned, the assessment may be restricted to non-intrusive techniques. However, if contamination is identified as a potentially significant liability issue, the client may consider an intrusive investigation necessary.

The form of the investigation will vary, depending upon the client's objectives and might include a geo-technical assessment of ground conditions to ascertain suitability for construction; sampling and analysis of soil, gas, ground and surface waters and other media to determine contamination levels; a hydro-geological assessment of groundwater flows; and a biological assessment to determine the presence of noxious or invasive plant species. If the site includes or is likely to include protected species or historical remains it will also be necessary to evaluate these and to adapt any sampling programme to avoid collateral damage.

Much of this information can be derived from the environmental assessment components outlined under ISO 14015. It should be noted however that ISO 14015 has a wider application than addressing site contamination. During the development of the standard several options were considered as the basis for the standard including the type of Phase I assessment often undertaken for contaminated sites. A decision was taken that the standard should be a hybrid of the Phase I assessment and the broader type of due diligence assessment normally used at operational sites. The intention was to avoid proliferation of standards and to provide a single document for use in several applications.

The various forms of intrusive investigation undertaken during environmental due diligence are each dependent upon the circumstances and objectives of the assessment. These include:

- pre-demolition audits for sites being decommissioned;
- site reports under the Pollution Prevention Control (PPC) Regulations 2000;
- ground investigations for sites being redeveloped or for Part IIA of the Environment Act 1990;
- monitoring of baseline conditions for operational sites during change of ownership;
- monitoring of baseline conditions for operational sites where information is needed either for Integrated Pollution Prevention Control (IPPC) applications, EMS, rehabilitation planning or other purposes.

Environmental Due Diligence

ISO 14015 is applicable to the preliminary assessment under each of these circumstances but the reader may refer to more detailed guidance from other sources for specific aspects such as:

- guidance for the design and implementation of site investigations including, for example, the ISO 10381 series for soil quality sampling and ISO 11464, *Soil quality – Pre-treatment of samples for physico-chemical analyses*;
- ground investigations for sites being redeveloped including BS 10175, *Code of practice for the investigation of potentially contaminated sites*;
- DEFRA and EA guidance for IPPC;
- methods for sampling and analyses of water, wastes and materials for which there are numerous ISO and other standards.

Remediation planning may form part of the environmental due diligence assessment but the implementation of remediation plans will not.

Sites being decommissioned

Sites being decommissioned may require two main types of investigation:

- pre-demolition audits;
- Surrender Site Reports (SSR).

Pre-demolition audits are currently limited in usage but are an extremely useful tool in restricting the spread of contamination across sites being decommissioned. The author has first-hand knowledge of numerous sites where uncontrolled demolition has resulted in widespread contamination that could have been avoided, or at the very least minimized and contained in specific areas.

The pre-demolition audit is undertaken prior to de-commissioning and identifies:

- the nature of on-site materials and wastes including, for example, asbestos, oils, treatment chemicals, leaded paint, fluorescent lighting, solvents;
- contaminated building materials such as flues or stacks with combustion residues, pipe-work or drainage containing waste streams. This also includes historic drainage, pipe-work and other systems that may no longer be in use but might still contain materials and wastes;
- sampling and analytical programmes to determine the nature of wastes and on-site materials;
- the presence of invasive plant species that can be spread if inappropriately handled. For example, it is illegal to spread Japanese knotweed under the Wildlife and Countryside Act 1981, and soil containing knotweed rhizomes must be handled in accordance with EA guidance when found on site. It can be expensive and time-consuming to remove and destroy knotweed;

Intrusive investigation

- re-usable clean materials such as clean hard-standing, brick and concrete that may be useful materials during site re-grading;
- areas to be used for segregating and containment of wastes and materials prior to recycling, treatment or disposal;
- legal requirements regarding treatment, handling, transport and disposal of wastes.

This type of audit is also an increasingly important aspect of waste minimization and increased control on the re-use of construction and demolition wastes.

Where a site licensed for specific industrial activity under IPPC is to be decommissioned and the IPPC permit surrendered, the regulator must be satisfied that the site has been returned to a satisfactory state before surrender. Here more detailed investigation will be needed.

Operators of Part IIA installations and mobile plants must report on site conditions at two stages. First, at the time of the application to operate through an Application Site Report (ASR), identifying baseline conditions and the preventative measures to be taken to avoid future pollution at the site. The preventative measures will include a Site Protection and Monitoring Programme (SPMP), possibly with a requirement to sample soil and groundwater conditions. The SPMP is expected to be proportionate to the risks posed by the site and may require an operator to collect reference data on substances (currently or in the future) used, produced, stored or transported under the permit. This reference data will be critical in judging the satisfactory state of the site on decommissioning.

The second is the Surrender Site Report (SSR) following decommissioning.

The ASR may be restricted to a non-intrusive assessment in accordance with ISO 14015. However it is in the interests of the operator to have as definitive an assessment as possible of the site condition in order that the SSR is limited to any adverse environmental change that is the liability of the operator. Consequently it may be in the best interests of the operators to carry out a detailed intrusive investigation, particularly where contamination is possible. Many IPPC regulated sites have a long industrial history and may occupy sites that are already contaminated. One possible adverse effect of the detailed investigation may be that the ASR may attract the threat of a Part IIA Remediation Notice.

The SSR requires the operator to demonstrate that the site has been returned to a satisfactory state and DEFRA's guidance [4] is that 'satisfactory state' is the pre-permit condition of the site. The SSR must prove that there has be no deterioration in site condition either by establishing that pollution prevention measures have been successful or, if any pollution has occurred, that this has been removed, treated or otherwise remediated. Technical Guidance Note IPPC H8 [5] indicates that operators should seek to:

- provide sufficient, relevant, reliable and unambiguous information to demonstrate that the site has not deteriorated as a result of permitted activities;
- if pollution has occurred, demonstrate that any additional pollutants have been dealt with in accordance with the PCC requirements.

Environmental Due Diligence

To assist, the Environment Agency (EA) provides an SSR report template. The SSR template includes the following broad headings:

- description of the installation showing relevant features and zones;
- operational activities undertaken at the site;
- substances handled including volumes and properties;
- reference conditions as given in the ASR;
- protective measures including:
  - SPMP and other monitoring records;
  - environmental data;
  - pollution incidents and mitigation;
  - site closure operations;
  - information on decommissioned areas;
  - surrender data;
  - schedule of reports;
- description of site condition at permit surrender;
- statement of satisfactory state.

Investigations for development sites

Whereas the focus of the SSR investigation is to demonstrate that there has been no additional contaminative burden imposed during the period of the permit, or if so, that this has been properly dealt with, many development sites have not been subject to IPPC control and little may be known about the condition of the site. Each contaminated site is unique, depending on past usage and the proposed end-use for the site. Before a commitment is given to any form of remediation, detailed investigations into the site's history and use need to be undertaken. See Figure 6 for a typical flow chart for the investigation and remediation.

Background Search
⇩
Site Survey
⇩
Risk Assessment
⇩
Remediation Plan

Figure 6 — A typical flow chart for the investigation and remediation

Intrusive investigation

If contaminated land is, for any reason, expected, there are a series of initial, relatively low cost activities which can be undertaken to confirm the situation. Sometimes referred to as a 'desktop' or 'Phase 1', these activities include the following.

- The identification of historic activities undertaken on site including industrial usage, waste disposal, landfilling or other potentially contaminating activities. This might be achieved through the examination of old site records, old street directories, historic maps, local authority planning and land use records; utility company records; archives and from discussions with local people. This will confirm the extent, duration and potential nature of any contaminating activities.
- The identification of any physical conditions such as ground instability due to historic mining or other voids or the presence of unsuitable ground conditions that may constrain development such as layers of peat.
- The identification of site geology and hydrogeology. This will help identify the potential of the substrata for trapping or assisting in the migration of contaminating materials.
- The identification of man-made site drainage and its ultimate destination. This will assist in the identification of any, perhaps off-site, reservoirs of potentially contaminated material.
- The examination of the site and surrounding area for 'sensitive receptors' – human residence, flora and fauna (especially protected species such as bats, badgers, etc.), waterways and cultural heritage.
- The identification of the existence of water boreholes or other abstractive processes.
- The completion of a physical site survey, identifying 'distressed' vegetation, evidence of fuel storage, asbestos and chemical usage, looking for evidence of 'leakage' of materials from the site, of any odours, soil discoloration or invasive plant species.

It will be noted that some elements of this type of assessment overlap with some of those of the pre-demolition audit although here the focus is on ground conditions and historic pollution.

If the evidence points to any potential for pollution, then a more intrusive investigation may be necessary, for which a conceptual model of the potential contamination will be necessary. This will include a risk assessment of the pollutants thought to be present and the hazards they potentially pose, the impacts the pollution might have on local receptors and the pathways by which they might migrate. The conclusion to this model might also consider the intervention methodology that might be necessary. The concept for this model is best summed up in Figure 7.

SOURCE ⇨	PATHWAY ⇨	RECEPTOR
Origin and nature of hazard	escape route/migration	potential for harm

Figure 7 — Conceptual model of potential contamination

Environmental Due Diligence

The use of risk assessment is discussed in more detail in chapter 9 but at its most basic there must be a linkage between these three components for there to be actual risk.

Intrusive investigation (Phase 2 investigation)

Once the decision has been made that contamination potentially exists, a more detailed intrusive investigation will be undertaken. It will be based around a statistically and practically sound approach developed through a conceptual model addressing the following issues:

- carefully assessed location, with evidence of the type and concentration of contaminants in soil and groundwater;
- identification, concentration and flow of gases;
- identification of soil characteristics – degree of porosity, permeability and leaching characteristics, geo-technical properties, pH value, water bearing strata and the provision of any evidence of organic matter.

Although intrusive investigation is the usual option, the reconnaissance of larger concentrations of contamination might be achieved using non-intrusive sampling of the following varieties.

- Microgravity, measuring changes in the gravity values arising from lateral and vertical density variations in the subsurface. Although this methodology can be used in areas where 'cultural' noise inhibits electromagnetic and seismic surveying, it is slow in creating data and its use in built-up areas is frustrated by local anomalies.
- Seismic refraction measures the compression or shear waves that have been refracted along an acoustic boundary and radiated back to the surface, where it is detected using an array of geophones. It can be used for detecting thickness and depth in lithological units with different densities and is especially useful in establishing the depth of the water table or identifying vertical boundaries such as old quarry walls. It is a slow process, often frustrated by 'cultural' noises and needs experienced operators to interpret the results.
- Infrared photography detects differences in reflected energy and can highlight distressed vegetation resulting from landfill gas or contaminated ground. It needs an airborne platform that provides variable quality photography due to difficulties in maintaining consistent heights and because of turbulence.
- Infrared thermography detects temperature differences in the ground that could be due to exothermic reactions in landfill sites or below-ground heating in coal rich soils. A helicopter or crane can be used, but the method relies on calm air conditions free from frost or snow.

In practice, these techniques are frequently expensive and confined to larger and more complex sites. The more traditional intrusive methodology include the following.

Intrusive investigation

- Driven tube samplers. These are portable and can assist with well-monitoring accepting a variety of measuring devices once a hole is formed.
- Hand auguring is portable and permits examination of soil profiles, although easier to use in sandy spoils.
- Spike holes allow cheap testing for ground gas and vapours but depth is limited.
- Trial pits and trenches formed using mechanical digging are cheap and permit a detailed examination of the ground condition. These are usually worked with a variety of power tools or vehicle mounted hammers and compressors. Boreholes can be driven using cable percussion or rotary drills.

The validity and accuracy of the information provided by the site investigation (SI) is critical. The Environment Agency provides guidance on QA/QC in relation to PPC SSRs, and the Technical Guidance Note H8 [5] identifies issues for consideration including:

- the rationale of the investigation and any constraints that may apply to its design or execution;
- the density/frequency of sampling locations and depths;
- the timing (including duration) or sampling and on-site testing;
- the methods used to carry out on-site testing or monitoring;
- the techniques used to collect, preserve, handle, store, ship (where laboratory analysis is carried out off-site) and prepare samples for laboratory analysis;
- the methods used to carry out laboratory analysis (noting that, from September 2004, soil analysis data that do not conform to the Agency's Monitoring Certification Scheme (MCERTS) will not be acceptable for regulatory purposes);
- the processing (e.g. statistical analysis) of data for interpretation and reporting purposes;
- the technical competence of individuals/companies involved in sampling, testing, analysis, data interpretation and reporting activities.

While much of this represents professional good practice in sampling and analysis and complies with ISO 14015 requirements that data should be validated as accurate, reliable, sufficient and appropriate with any constraints identified, what is new is the emphasis on MCERTS. For some time there has been concern regarding the interpretation based on laboratory analytical results, particularly where this is done without awareness of the constraints of different methods of analysis. This is particularly true of organic contaminants that can be influenced by the extraction media as well as the analytical technique used.

To take an example, Total Petroleum Hydrocarbons (TPH) are frequently analysed but what laboratories report is so variable that care must be taken in the use of the results. Some common problems are results that are not comparable because:

- one set is expressed as dry weight and another as wet weight;
- one set is from soxhlet extraction while another is not – the extraction method will extract some hydrocarbons but miss others.

Environmental Due Diligence

Furthermore, petroleum characteristics vary and some are more toxic than others. Crude oils differ from distillates and reporting both as total TPH is misleading as to the true environmental effects.

MCERTS is an accreditation scheme set up by the Environment Agency with the intention of improving the quality of monitoring data and uses the following standards as a basis for QA:

- ISO 17025, for monitoring and equipment testing;
- EN 45004, for inspection;
- EN45011, for product certification;
- EN 45013, for personnel competency.

ISO 17025:2000 is the performance standard for chemical testing of soil and covers (see www.environment-agency.gov.uk):

- the selection and validation of methods;
- sampling pre-treatment and preparation;
- the estimation of measurement uncertainty;
- participation in proficiency testing schemes;
- reporting of results and information.

The BSI website (www.bsi-global.com) provides information on these standards, and the EA website (www.environment-agency.gov.uk) on MCERTS.

Other investigations

Investigations for other purposes such as establishing baseline conditions for ASR, EMS, rehabilitation or liability planning will comprise elements of both previous types of assessment but are likely to include additional elements such as:

- noise and vibration;
- air quality;
- effluent and/or water quality;
- ecological assessment;
- archaeological or cultural heritage assessment;
- health and safety;
- product life cycle assessment.

In some cases, for example where assessments are for UK investors operating in developing countries, social impact assessments are likely to be a critical component with potential risks, especially risk to reputation, that may exceed environmental concerns. Assessment components may include:

- child and forced labour;
- involuntary resettlement;
- community consultation;
- HIV/AIDS and other health risks.

Guidance on these types of assessment is provided by the World Bank, IFC, UNEP and other organizations.

Conclusions

The form of intrusive investigation is dependent upon the circumstances of the investigation, the client's objectives and the applicable regulatory regime. Whilst there is some overlap in site investigation components there are elements in common including:

- the need for conformance with good practice and rigour in the assessment process;
- the need for demonstrable and, if necessary, auditable reliability and sufficiency of data.

Chapter 9 – Risk assessment and remediation

Risk assessment

Environmental Risk Assessment (ERA) is essential particularly in relation to issues of land contamination and is a requirement of EPA Part IIA. There are various methods and guidance documents for risk assessment in relation to soil, water, human health and ecosystems.

There has been a move away from the application of threshold and trigger levels as developed by the Interdepartmental Committee for the Redevelopment of Contaminated Land (ICRCL) that were used as the basis of decisions for remediation requirements for many years. The ICRCL values were withdrawn in 2002 and have been replaced by the Contaminated Land Exposure Assessment Model (CLEA) and Contaminated Land Reports (CLR) which uses Soil Guideline Values (SGVs) and health criteria values.

CLEA is a human health risk assessment tool which:

- is a computer model for assessing long-term risks for human health;
- uses the source–pathway–receptor risk approach;
- provides the basis for SGVs.

SGVs are available for a number of contaminants including arsenic, cadmium, mercury, chromium, nickel, lead and selenium and there are reports for a range of organics including benzo-a-pyrene (BaP), phenol, toluene, ethylbenzene, inorganic cyanides, dioxins and polychlorinated biphenyls (PCB).

CLEA uses receptor and exposure pathway, exposure time, soil type, land use, ambient exposure and toxicological properties to provide SGVs in relation to long-term human health risk. CLEA does not consider certain land uses, short-term risk (with the exception of cyanide) and does not consider controlled waters or ecosystems as receptors. Nor does it address worker exposure during site work and redevelopment. Consequently there are limitations in the application of CLEA and circumstances where alternative risk assessment models should be used.

At present, CLEA is in abeyance whilst work is undertaken on some of the underlying assumptions and on SGVs for more determinants.

In circumstances where CLEA is not appropriate, the following methods are internationally accepted: RBCA, SNIFFER, ConSim and RISC. These are software models (or a spreadsheet in the case of SNIFFER) for soil and human health risk or (in the case of ConSim) for groundwater receptors.

Health criteria supplied by the UK Health and Safety Executive (HSE), World Health Organization (WHO), the US Environmental Protection Agency (EPA) and others may be appropriate for use.

Environmental Due Diligence

Regulation for contaminated land frequently refers to 'significant risk' or 'significant harm' without defining 'significant'. To simplify the assessment of risks to determine that significant harm is being caused, the Environmental Protection Act 1990, Part IIA allows local authorities to use scientifically based guideline values such as those generated by CLEA in making decisions as to the presence of contamination and the need for any remediation action.

Figure 8 (taken from DEFRA Guidance) shows the process for general environmental risk assessment (ERA).

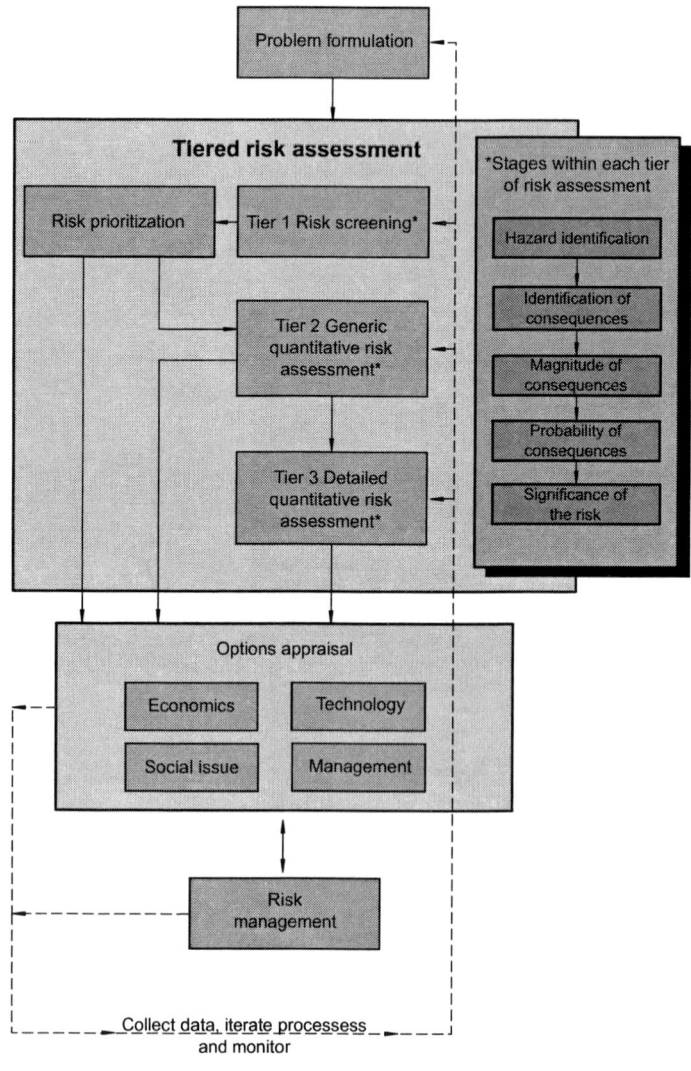

© DEFRA 2000

Figure 8 — General risk assessment process

Risk assessment and remediation

The EA is developing guidance on a framework and methods for assessing harm to ecosystems from contaminants in soil. The Eco-risk assessment [6] is intended to apply to any ecosystem or living organism that is afforded legal or policy protection including:

- sites of special scientific interest (SSSI);
- national nature reserves;
- marine nature reserves;
- bird protection areas;
- habitats covered by Planning Policy Guidance Note 9;
- nature reserves under the National Parks and Access to the Countryside Act 1949;
- any European site within the meaning of Regulation 10 of the Conservation (Natural Habitats etc.) Regulations 1994;
- any candidate Special Areas of Conservation or potential Special Areas of Conservation given equivalent protection.

The risk assessment procedure will include Soil Screening Values (SSVs) for ecological protection similar to the SGVs that are derived for protection of human health.

Data for use in ecological risk assessment should include:

- chemical data;
- biological data (e.g. invertebrate testing);
- ecological survey on extent, diversity and conditions of populations.

The risk assessment guidance provided by both CLEA and the up-coming 'ecorisk' assessment are anticipated to apply to the EU Liability Directive as discussed in chapter 10.

Risks to controlled waters are encompassed by EA Research and Development Document 20 [7]. This provides guidance on assessment methods to derive the level of remediation required to protect ground and surface waters. It forms part of the overall process to evaluate health and environmental risks that contaminated soil and groundwater presents. The methodology is based on source–pathway–receptor analysis that leads to the derivation of site-specific remediation criteria based on the potential impact at the identified receptor.

Remediation

Once contamination has been identified, a variety of options for remediation are available. They fall into five main categories:

- physico-chemical:
  - solidification/stabilization;
  - soil vapour extraction;
  - soil washing;

- biological:
  - windrows;
  - land farming;
  - bio-piling;
  - composting;
  - natural attenuation;
- thermal:
  - thermal desorption;
  - pyrolysis;
- excavation and disposal;
- engineering/containment:
  - cut-off walls;
  - capping/impermeable membranes;
  - gas venting systems;
  - sub-surface drains;
  - groundwater pumping.

Of these categories, the last two are not true remediation techniques since they either transfer the contamination elsewhere or seek to contain it.

It is not proposed to discuss remediation techniques in detail here but instead to consider current remediation practice, how this is developing under recent legislation and what the implications of this are for environmental assessment of contaminated sites. However, Appendix 1 contains information on several remediation techniques.

Traditionally in the UK 'dig and dump' and replacement with clean fill has been the favoured practice for removal of contaminated soils. This is largely because much of the UK's contaminated soil originates from small development sites where pressures for removal, time and space constraints and limited volumes render on-site treatment technologies uneconomic. Bio-remediation has been used for hydrocarbon contaminated soils at larger sites such as former petroleum or coal-gas sites. Engineering or containment systems are also used, particularly where gas or leachate is a constraint.

There are several pieces of environmental legislation that have affected and will continue to affect the use of brownfield land and legal environmental liabilities associated with contamination. With the introduction of the Environmental Protection Act in 1990 it was expected that regulation would stabilize and consolidate. However, the period since has been one of continued uncertainty; the regulatory framework is still developing and there is real and sustained uncertainty in the marketplace. From the point of view of the Local Authorities and the Agency, Part IIA priority sites are being identified and the Environmental Protection Act 1990 is beginning to bite as shown in the recent Circular Facilities Case (see Case study 1).

Risk assessment and remediation

Case study 1 – Part IIA Contaminated Land Regime
Circular Facilities (London) vs Sevenoaks District Council (Kent)

The case concerned a site previously used as a brickworks within the boundaries of Sevenoaks District Council. Following the closure of industrial operations on the site, clay pits filled with water, silt and vegetation. The site was purchased in 1978, and subsequently bought in a merger between a Mr Scott (landowner) and Circular Facilities (developer). During the 1980s eight residential properties were constructed on the site, and sold to private residents. In 1991, the site was identified as a landfill gas hazard due to the potential danger of sub-terrestrial bio-generation of methane and carbon dioxide, prompting the District Council to undertake gas protection measures at their own expense.

Some years later, following the introduction of the above statutory legislation, Sevenoaks District Council commissioned a report investigating whether the gas posed a significant possibility of significant harm to the residents' health. The report concluded that such a possibility existed, prompting Sevenoaks District Council to deem the land as contaminated, and serve a remediation notice to Circular Facilities requiring the venting of the landfill, and mitigation against anaerobic methane production. Circular Facilities appealed against this notice, at which point the case entered the magistrates court. Section 78N of the Environmental Protection Act 1990, enabling authorities to undertake works themselves and later seek to recover costs from liable parties, was exercised by Sevenoaks District Council, which conducted £46 000 of work.

The minor appeal by Circular Facilities was rejected by the magistrates court on a further ruling of section 78N of the Environmental Protection Act 1990, leaving the more substantive issue concerning the reasonable identification of Circular Facilities as liable. A liable party is one that 'caused or knowingly permitted' the presence of the substances which led to a site being formally determined as contaminated. The contaminating substances can be either a secondary pollutant formed from a chemical reaction or, as in this instance, the primary pollutant caused by a biological process.

A geo-technical report performed on the site pre-development was produced as evidence, detailing that organic matter and product gases had been observed in the clay pits in the site. The presiding judge concluded that Circular Facilities must have been privy to this information, considered the risk of gas production, yet continued their development. Therefore, they had knowingly permitted the presence of primary pollutants from a biological process, causing contamination.

For land to be deemed contaminated, a significant pollution linkage must exist between contaminants and receptors. Statutory guidance under Part IIA enforces six exclusion tests for potentially liable parties, the sixth and final test being of most relevance to this case. This test excludes the person who permitted the presence of pollutants, if another person established the pollution linkage. The judge concluded that the owners previous to Circular Facilities who had introduced the pollutants were not liable, as Circular Facilities had introduced the pollutant linkage. The judge dismissed the appeal request and ordered £15 000 in costs to be paid to Sevenoaks District Council.

However the major concern remains the implementation of the Landfill Directive 1999, which came into force in July 2002 and has far-reaching implications for contaminated land.

Environmental Due Diligence

The Landfill Directive and its supporting legislation redefines and increases the types of material considered hazardous (and some soils and materials from contaminated sites fall into this category), requires the treatment of hazardous wastes prior to landfill and bans co-disposal of hazardous and non-hazardous wastes. Furthermore, this coincides with government proposals to:

- increase the re-use of brownfield land, setting targets of 60% of new build to be on brownfield sites;
- reduce the volume of material going to landfill by 85%.

Two aspects of the Landfill Directive will affect the capacity for, and costs of, disposal of contaminated soil. First is the reduction in available landfill space to accept hazardous waste and second, the need to treat such wastes prior to disposal. According to a report from the Landfill Directive and Regeneration Task Group of possible impacts and mitigation options of the landfill directive in 2003, of more than 200 landfills that previously accepted hazardous wastes, only 37 sites will now be available under the new regulation of which only 9–13 will be commercially available sites. Moreover, these sites are unevenly distributed and some regions such as the South East, the South West and Wales will have limited hazardous waste landfill capacity.

The constraints on 'dig and dump' are anticipated to result in a temporary slowdown in regeneration of brownfield land whilst alternatives are established. There are several options for treatment and the technology is not new although some is comparatively untried in the UK. The principal issues for most developers are the need for rapid removal of contaminated soil and materials off-site and the costs associated with treatment. English Partnerships is quoted as anticipating an increased remediation cost per hectare ranging between £11 000 and £500 000, with an average increase of £50 000 for large sites in its National Coalfields Programme. The costs result from increased landfill gate prices, haulage costs and the potential costs of pre-treatment. The situation for small sites, the most common situation for new housing development, is anticipated by the House Builder's Federation to add an average of £2000 per housing plot to development costs. Small sites also have little or no opportunity for on-site treatment.

This situation has both positive and negative impacts. The economic impacts and development constraints are clear and there are additional impacts such as increased transportation costs and the risks that some sites will no longer be economically viable for remediation and may be left as eyesores. However 'dig and dump' is not sustainable and improved waste management provides environmental benefits.

In response to this a feasibility study that was sponsored by the Soil and Groundwater Technology Association examined the possibility of using temporary soil treatment centres known as Hubs where contaminated soils transferred from numerous small sites will be treated and the resultant clean material either returned to site or sold as fill. At the moment the concept is new and will require reform of some of the anomalies and constraints that currently bedevil brownfield regeneration over and above the issues of landfilling including:

Risk assessment and remediation

- the need for a single Remediation Permit system for contaminated land to replace and streamline the current complex system whereby a site may need a Waste Management License that is time consuming to obtain, a mobile plant license for treatment or may be granted an exemption by the regulator. DEFRA was undertaking a Waste Permitting Review to address the existing system but abandoned this in August 2004;
- that post-treated soils should not be classified as waste.

Such Hubs are used successfully in Europe where dedicated permitting is in place and distinctions are made between disposal and re-use of 'waste' that removes some of the blight associated with this.

Conclusions

To summarize, the following conclusions have been made.

- Risk assessment is an integral part of any environmental assessment and is extending beyond the traditional assessments of human health risks to include ecological risks.
- Guidance on methods of risk assessment is in continued development and will require more specialist skills including toxicology and ecology.
- Remediation can no longer rely on 'dig and dump' as a solution to ground contamination and will need to be replaced by on-site treatment where this is possible or off-site treatment methods for smaller sites.
- Site remediation is becoming increasingly complex and more expensive whilst uncertainty regarding remediation permitting, waste licensing and other issues is set to continue.

Chapter 10 – Emerging legislation

The Liability Directive

So far the discussion has focused on site clean-up; however, there is pending legislation under the EU Liability Directive to extend liability to damage to natural resources. To a degree this follows the system in the US where polluters are held liable for several successive types of clean-up including:

- removal of contamination;
- primary restoration of natural resources such as habitats, water, flora and fauna and amenity use to baseline conditions;
- compensatory restoration to compensate for the loss of ecological and other services for the period between the onset of damage and the restoration to baseline conditions. This might be done by, for example, the establishment of a new nature reserve or additional habitat to provide the same services such as bird feeding and nesting grounds.

Natural Resource Damage Assessment (NRDA) is determined in the US by guidance provided by the Department of the Interior and the National Oceanic and Atmospheric Administration within the US Department of Commerce under the Comprehensive Environmental Response, Compensation and Liability Act 1980 (CERCLA or Superfund) and the Oil Pollution Act 1990. The vagaries of Superfund are too well known to be repeated here and are only introduced to provide context to the EU Liability Directive and the potential implications for additional remediation requirements.

The EU Liability Directive came into force in April 2004 and is expected to be implemented in Member States by April 2007.

Damage covered by the Liability Directive restricts damage to:

- biodiversity protected at community and national levels (specifically under the Habitats and Wild Birds Directives);
- water covered by the Water Framework Directive;
- human health when the threat to health is land contamination.

The European approach on NRDA is somewhat different from that of the US. The EU Liability Directive requires that NRDA should apply only where damage is to environmental media with legal protection or where health is threatened and this is deemed to be significant. The Liability Directive is not retrospective, only applying to new damage and requires that all sums recovered must be used on restoration. Unlike CERCLA, the EU Directive gives an explicit preference to least cost options and relies on costs of restoration rather than monetary valuation of resources.

Environmental Due Diligence

According to Directive 2004/35/CE of the European Parliament and of the Council of 21 April 2004 on environmental liability with regard to the prevention and remedying of environmental damage, in respect of biodiversity, the significance of damage is that which 'has adverse effects on reaching or maintaining the favourable conservation status of habitats or species has to be assessed by reference to their conservation status at the time of the damage, the services provided by the amenities they produce and their capacity for natural regeneration'.

Remedying damage is achieved through 'the restoration of the environment to its baseline condition by way of primary, complementary and compensatory remediation, where:

- 'Primary' remediation is any remedial measure which returns the damaged natural resources and/or impaired services to, or towards, baseline condition;
- 'Complementary' remediation is any remedial measure taken in relation to natural resources and/or services to compensate for the fact that primary remediation does not result in fully restoring the damaged natural resources and/or services;
- 'Compensatory' remediation is any action taken to compensate for interim losses of natural resources and/or services that occur from the date of damage occurring until primary remediation has achieved its full effect;
- 'Interim losses' mean losses which result from the fact that the damaged natural resources and/or services are not able to perform their ecological functions or provide services to other natural resources or to the public until the primary or complementary measures have taken effect. It does not consist of financial compensation to members of the public.

Where primary remediation does not result in the restoration of the environment to its baseline condition, then complementary remediation will be undertaken. In addition compensatory remediation will be undertaken to compensate for the interim losses.'

For the purpose of remediation:

'Complementary remediation is to provide a similar level of natural resources and/or services including, as appropriate, at an alternative site, as would have been provided if the damaged site had been returned to its baseline condition. Where possible and appropriate the alternative site should be geographically linked to the damaged site, taking into account the interests of the affected population.

Compensatory remediation shall be undertaken to compensate for the interim loss of natural resources and services pending recovery. This compensation consists of additional improvements to protected natural habitats and species or water at either the damaged site or at an alternative site.'

It remains to be seen as to how Directive 2004/35/CE will be implemented and what the scale of associated liability will be.

Other legislation

Whilst the Liability Directive and the recent amendments following the Landfill Directive are of most immediate concern in respect of due diligence in the context of contaminated land, there is other legislation either being implemented or about to be implemented that will impact on environmental due diligence assessments including:

- the Waste Electrical and Electronic Equipment (WEEE) Directive and Restriction of Hazardous Substances in Electrical and Electronic Equipment (ROHS) Directive that will affect the manufacture and disposal of electrical and electronic equipment;
- the Water Framework Directive which will affect all aspects of water and water management especially sewage treatment and effluent discharges. Virtually all industry will be affected, particularly those with high water usage, and water pricing policies will be introduced to provide incentives to reduce water usage. The Directive will also introduce a catchment/river basin approach to water management;
- the Solvent Emissions Directive which will affect a range of industries including paints, textiles, printing and pharmaceuticals;
- the Packaging and Packaging Waste Directive;
- the Chemical White Paper is intended to stimulate safety and innovation in the use or replacement of chemicals.

It will also be interesting to see what evolves from EU Council Decisions on climate change, greenhouse gas emissions and emissions trading and effects on energy usage.

Conclusions

Environmental legislation, particularly that emanating from the EU, will be a contining driver for environmental change and any environmental due diligence assessment will need to address existing regulations and anticipate the effects of any future regulations on reputation or finances and operations. The evaluation of these business consequences may well be the most important feature of the assessment.

ISO 14015 may have the components in place to address many of the issues raised here but it is already evident that the review of the standard scheduled to take place in 2006 will take into account rapidly changing perspectives in environmental management and most especially in the protection of ecosystems, use of raw materials and other natural resources and dealing with wastes safely. These issues are already covered by the standard but will need to be made more explicit, possibly with supporting guidance in the form of case studies.

We have not discussed social issues or social corporate reporting here. Public consultation is already a feature of many aspects of environmental assessment and issues of governance and ethical performance are affecting many high profile sectors. It is highly likely that these aspects too will be discussed during the review.

Environmental Due Diligence

Similarly, environmental and health and safety look set to converge yet further. In some large-scale environmental due diligence assessments such as those undertaken for the mining sector, health and safety issues can form the largest liabilities. We have already seen the alignment of environmental and quality auditing and pressures will continue to align health and safety in a similar manner.

The intention here is not to pre-judge the review, merely to anticipate the legal and other pressures that may underpin reform of the standard.

Appendix 1 – Remediation technology

Introduction

Remediation techniques fall broadly into the following categories (see chapter 9, Risk and remediation):

- physico-chemical:
  - solidification/stabilization;
  - soil vapour extraction;
  - soil washing;
- biological:
  - windrows;
  - land farming;
  - bio-piling;
  - composting;
  - natural attenuation;
- thermal:
  - thermal desorption;
  - pyrolysis;
- excavation and disposal;
- engineering/containment:
  - cut-off walls;
  - capping/impermeable membranes;
  - gas venting systems;
  - sub-surface drains;
  - groundwater pumping.

The choice of an appropriate technique will depend upon a variety of issues, among the most important of which are the following:

- contaminating material, including type, toxicity, concentration;
- depth and extent of contamination;
- soil quality, including the presence and concentration of bacteria, existence of nutrients, soil neutrality and moisture content;
- geographical/physical location;
- meteorological conditions;
- geological and hydrogeological conditions;
- potential hazards arising from the presence of a contaminant;

- pathways to sensitive receptors;
- existence of groundwater and groundwater extraction points.

In considering techniques for remediation, time is important, especially where (re)development is intended and where money (usually in the form of working capital costs) is a consideration. Before examining the remediation techniques, it is worth reviewing the issues identified (see list immediately above), in order to gain an insight into the factors influencing the choice of technique.

Contaminant

These are often described generically as dense non-aqueous phase liquids (DNAPLs) or low-density non-aqueous phase liquids (LNAPLs). The more dense material would typically reflect the 'heavier end of the barrel', for example, lubricating oils, bitumens, tars and fuel oils. These are difficult to volatilize and may almost certainly require the application of steam or some other form of heating to assist in their dispersal. The lighter material would typically comprise volatile organic compounds – hydrocarbon material such as petroleum, diesel and gas oil and solvents – which are more volatile, and will almost certainly require some form of containment or capture to limit their escape to atmosphere.

The DNAPLs of higher molecular weights are difficult to break down during land farming operations, and are also slow to biodegrade. LNAPLs can be more readily biodegradable and can be removed through land farming operations, subject to local regulations on the release of volatile organic material.

The presence of toxic material or heavy metals will require separate, additional treatment to ensure their recovery or total destruction.

Depth and extent of contamination

The deeper the material has sunk, the wider the spread (the plume) of the material. In the case of DNAPLs, this can be troublesome given the extent to which the effectiveness of *in situ* heat or steam treatment can be dissipated. Extraction of the material for heat treatment is an alternative, whilst the interference in any potential pathways for migration of the material is another.

Volatile materials at a shallow depth can be readily addressed through sparging or soil vapour extraction, although there is a Radius of Influence (ROI) that relates to the general 'envelope' in which the technologies will function effectively.

Natural attenuation may be a more sustainable and cost-effective alternative for deeper-lying material.

Soil quality

Soil chemistry is a topic in its own right and it would be impossible to do the subject complete justice in a few short paragraphs. For the purposes of this book, it is probably appropriate to focus on the following critical elements with regards to soil quality.

- *Bacteria.* Their characteristics vary enormously, but the presence of bacteria is vital to the encouragement of the decomposition (biodegradation) of contaminants, whether they are being treated *in situ* or have been extracted for some form of composting or land farming.
- *Moisture content.* Soil must be neither too saturated nor too dry. Dry circumstances will discourage the effectiveness of biodegradability, while saturated soil will inhibit volatilization and, hence, the efficiency of several of the remediation techniques.
- *Soil nutrients.* Bacteria require nutrients, especially nitrogen and phosphorus, for cell growth, and the presence of carbon is essential for that purpose and for the energy to sustain metabolic functions associated with growth.

Geographical/physical location

The geographical location of the contamination may make access for some mobile remediation technologies difficult. Similarly, the presence of contaminants in confined locations may be physically limiting where intrusive techniques are desirable for the removal of the contaminant. The costs associated with transport may also make some techniques less competitive.

Meteorological conditions

Temperature and wind conditions can have a substantial impact on the chosen treatments. Cold temperature discourages volatilization and inhibits biodegradation. Warm temperatures have the reverse effect, although in dry climates biodegradation will be slowed in the absence of moisture.

Wind direction and strength may be a limiting factor if venting or composting, entailing regular turning of soils, are to be considered. Special covers or barriers may need to be provided to control the deposition of soil, dust or vapours away from the treatment site.

Geological/hydrogeological conditions

The ease with which probes and venting systems can be installed will be largely dictated by the depth of the contamination and the porosity of the underlying rock and soil at the site. Drilling for access increases costs.

In the presence of groundwater, techniques involving some form of air blowing or 'sparging', or any techniques involving air pressure, may be limited in their application. This is due to the threat of displacement of the groundwater with consequential impacts on the profile of the contamination, or the potential for interference in abstraction processes.

Potential hazards from the contaminant and potential contaminant pathways

Many of the software programmes for evaluating risk have already been identified in chapter 9. However, the identification of the sensitive potential receptors may require an intensive reconnaissance of the area to determine whether humans, flora and/or fauna are present. The ways (e.g. subterranean and airborne) in which contamination can migrate may also require air movement and air quality modelling, and some detailed and probably intrusive reconnaissance of the geology and hydrogeology, in order to determine potential migratory routes and the potential for physical intervention.

Groundwater and groundwater extraction

The presence of groundwater or groundwater extraction points near the source of contamination will influence the choice of intrusive intervention. The presence of groundwater close to the surface will limit the effectiveness of some remediation techniques. Over-pressurization of lower levels of groundwater may impair the effectiveness of the remediation process or threaten the abstraction process.

The physico-chemical, biological and thermal remediation techniques are described in the following.

Physico-chemical

- *Solidification/stabilization*. As the name implies, this technique, using concrete or, in extreme cases, glass, immobilizes the contamination but does not destroy it. It is an expensive treatment process, in the region of £300 to £400 per cubic metre.
- *Soil vapour extraction*. A process whereby wells are sunk into the ground adjacent to the source of contamination, and a small vacuum is drawn. Light, volatile materials are encouraged to vaporize and can then be captured through the gallery of wells for treatment. This is a commonplace and fairly economical treatment using mobile equipment, but may take 6 months to 12 months to be effective.
- *Soil washing*. This technique uses bulk washing techniques to separate contaminant particles before using some chemically-enhanced water to separate the contaminant for recovery. A cheap and commonplace treatment.

Remediation technology

Biological

- *Windrows*. Contaminated soil is thrown up into rows, perhaps 1.5 m to 2 m high, and allowed to 'weather'. Aerobic decomposition, essentially, takes place – moisture and reasonably high ambient temperatures are also pre-requisite conditions. This technique is energy consuming in that the windrows require frequent turning, usually by some form of mechanical process. Shelters or barriers may be required to reduce airborne movement, and drainage to capture leachate and stormwater run-off is important. The process is relatively cheap, but requires access to spare land and may take 12 months or more to be effective.
- *Land farming*. A similar process to that above, but occupying larger areas of land and requiring a 'tilling' or rotary ploughing process. Protection measures and efficaciousness are similar.
- *Bio-piling*. Also similar to windrowing and land farming, although the material is piled up to between 1.5 m and 3.0 m and usually uses forced or injected air to accomplish the microbial decomposition.
- *Composting*. May use anaerobic processes to assist decomposition and will require the generation of quite high temperatures in the compost pile to generate the destruction pathogens, seeds and other materials. Costs of all these techniques are between £20 and £60 per tonne and could take 12 months or more to be effective.
- *Natural attenuation*. Contamination is left in place, perhaps after
- *Reduction in concentration, and monitoring its decomposition*. In some cases, the pathway to any sensitive receptors may be blocked. The costs of monitoring for decomposition or any migration are usually much lower than traditional methods of remediation and the process is deemed to be a more sustainable process for managing contamination.

Thermal

- *Thermal desorption*. This is an *in situ* process where heat is applied to the contamination to separate hydrocarbons from the soil, and capture the ensuing vapour for treatment, usually catalytic oxidation or carbon absorption. It is an effective process for dealing with contaminants of low molecular weights with hydrocarbons, is mobile and useful for treating quite large volumes of material at a cost of up to £70 per tonne.
- *Pyrolysis*. An 'ex-situ' process, which involves the 'baking' of the contaminated material to permit the evaporation and neutralization of the hot gases before release. The process is inhibited if the soil is saturated in water. An energy intensive process but capable of dealing with quite large volumes of soil at a competitive price.

Excavation and disposal

'Dig and dump', once a traditional and fairly cheap method of disposal, is now discouraged through changes in the regulations governing the disposal of hazardous waste to landfill, and the unsustainable nature of the process. May still be permissible in countries outside the European Community and the US.

Engineering/containment

Examples of types of engineering/containment include the following.

- *Cut-off walls*. As the title implies, the construction of walls to contain the contamination.
- *Capping/impermeable membranes*. Used in locations where some residual contamination remains on site and is designed to prohibit any flows of gas into foundations or premises.
- *Gas venting systems*. Galleries of collection lines are left *in situ* to collect residual gas for treatment. Most frequent applications occur in completed landfill sites, where methane is captured for use as a renewable energy source and powers small capacity reciprocating kits, or gas turbine generating kits.
- *Sub-surface drains*. Manufactured drainage to ensure that any contaminated material leaching from a site may be captured and pumped away for treatment.
- *Groundwater pumping*. Designed to lift groundwater in the zone of previous or residual contamination to remove or monitor water quality.

References

[1] ISO 14004:2004, *Environmental management systems – General guidelines on principles, systems and support techniques*.

[2] ISO 19011:2002, *Guidelines for quality and/or environmental management systems auditing*.

[3] ISO 14015:2001, *Environmental management – Environmental assessment of sites and organizations (EASO)*.

[4] DEFRA. *Integrated Pollution Prevention and Control: A Practical Guide*. 3rd Edition. UK: DEFRA; 2004.

[5] Environment Agency (EA). Technical Guidance Note IPPC H8. Guidance on the Protection of Land under the PPC Regime: Surrender Site Report. Consultation Draft Version 1.0. Bristol, UK: EA; 2004.

[6] Environment Agency (EA). Ecological Risk Assessment. A public consultation on a framework and methods for assessing harm to ecosystems from contaminants in soil. Bristol, UK: EA; 2003.

[7] Environment Agency (EA). Research and Development Document 20. Methodology for the derivation of remedial targets for soil and groundwater to protect water resources. Bristol, UK: EA; 1999.